A Probabilistic Model of the Genotype/Phenotype Relationship: Does Life Play the Dice?

If, in some cataclysm, all scientific knowledge were to be destroyed, and only one sentence passed on to the next generations of creatures, what statement would contain the most information in the fewest words? I believe it is the atomic hypothesis (or the atomic fact, or whatever you wish to call it) that all things are made of atoms—little particles that move around in perpetual motion, attracting each other when they are a little distance apart, but repelling upon being squeezed into one another. In that one sentence, you will see, there is an enormous amount of information about the world, if just a little imagination and thinking are applied.

Richard Feynman

A Probabilistic Model of the Genotype/Phenotype Relationship: Does Life Play the Dice?

Jean-Pierre Hugot
Professor
Université Paris Diderot Sobonne Paris-Cité
Paris, France
and
Head
Department of Pediatric Gastroenterology
Robert Debré Hospital
Paris, France

Translated by Clare Davis

CRC Press
Taylor & Francis Group
Boca Raton London New York

CRC Press is an imprint of the
Taylor & Francis Group, an **informa** business
A SCIENCE PUBLISHERS BOOK

Translation of: Comment dessiner un mouton? Essai sur la relation genotype/phenotype Les Editions du Net, Paris 2015.

CRC Press
Taylor & Francis Group
6000 Broken Sound Parkway NW, Suite 300
Boca Raton, FL 33487-2742

Printed on acid-free paper
Version Date: 20180420

International Standard Book Number-13: 978-1-138-32072-7 (Hardback)

Visit the Taylor & Francis Web site at
http://www.taylorandfrancis.com

and the CRC Press Web site at
http://www.crcpress.com

To

Marion, Mahaut and Guilmot
for their forbearance with my
constant condition of researcher

and

Jean-Marc
Research companion
and friend
without our tireless discussions
this book would not exist

Preface

A fundamental question in biology is the relationship between the genome and the organism, in other words, between the genotype and the phenotype. How does a collection of genes produce a given individual? Up to now, this question has been considered as a problem which will be resolved by the analysis of the role of each gene. Molecular biologists can now dissect the biological function of genetic variations, one by one, and in the end, it is expected that it will be possible to predict the organism knowing its genetic makeup. The puzzle is complex but the full image will be visible when all the pieces have been assembled.

Unfortunately, it appears that even with an accurate knowledge of the genome, it is difficult to predict some traits like complex genetic disorders. This limit questions the current paradigm about the genotype/phenotype relationship. It is thus necessary to explore alternative avenues to propose new hypotheses. This work explores a new kind of relationship between the genotype and the phenotype where the genotype provides only a panel of possible phenotypes without strict determinism.

This book is the translation of a book published in French in 2015 under the title "Comment dessiner un mouton?" which could be translated as "How do you draw a sheep?". This title makes reference to the book by Antoine de Saint Exupery "the little Prince". In Chapter 2, the little Prince meets Saint Exupery and asks him: "draw me a sheep". The author draws several cartoons but none of them is judged convenient. At the end, he draws a box with few holes and says: "this is the box, your sheep is inside".

The main message of the present book is well illustrated by this story. As Saint Exupery, the genome does not decide what is in the box. At best it gives the indication that there is a sheep inside the box but nobody can predict which sheep it is. The sheep takes any shape or size or characteristics on its own, only limited by different limitations such as the size of the box and... the wishes of the little Prince.

Jean-Pierre Hugot

Contents

PART 3: EVOLUTION

Warning

As a paediatrician and researcher, I specialised in Crohn's disease, a disabling disease characterised by chronic diarrhea and an alteration of well being. My reflection on this illness and complex genetic disorders in general have fuelled my approach and serve as an introduction to this work. However, this book is not directly linked to my work as a doctor and researcher. It has only a light bearing on diseases but on the contrary questions other biological problems of a more general nature. As a result the reader may deduce that I am not an expert in the multiple fields broached herein. This work is the attempt of an amateur to be enlightened. Consequently, certain mistakes or misinterpretations may be noted by specialists. For this I apologize in advance and hope they will nevertheless find some interest in these pages.

The work is not scientific in the usual sense of the word by research biologists, professional category to which I belong. Thus, it does not present a hypothesis, an experimental plan and a discussion of original results from experiments. On the contrary, it is based on ideas and concepts. The book should therefore be taken as a reflection. On the other hand, it talks about biology and is based on observed data. Hypotheses and models are generated which can be tested by the experimental method. It therefore has a link with science. I would willingly give it the slightly old-fashioned term of 'Essay on natural philosophy' to illustrate the link with a deductive rather than experimental approach.

As the reader will discover, the ideas developed in this book feed on analogies between the concepts of quantum physics and those of biology. An analogic approach has been used since it is an effective method for reflection. It is nevertheless only a tool. It is not an ideological choice, and it is only for its operational qualities simply that the analogical approach has been used. Like the microscope, it provides an original view point to look at the living. It is in no way to research or promote any mystical or esoteric thought as may have been done by others.

Introduction: Complex Genetic Diseases

No thought is more important than that which focusses on the transmissibility of aptitude and character.

Nietzsche

In the simple model corresponding to Mendel's Laws and classic genetics, the correlation between the genotype and the phenotype is perfect.[1] In other words, a given genotype corresponds to a given phenotype and vice versa. A pea is smooth or wrinkled according to the existence of a molecular shape (called allele)[2] or the other of the gene. The relationship is so unequivocal that the allele, since Morgan, carries the name of the phenotypical trait "smooth" or "wrinkled" whereas it expresses in reality only information on the sequence of the DNA. This direct and symmetric relationship [genotype ↔ phenotype], allows reliable and simple predictions. They are expressed by Mendel's laws for the transmission of phenotype from one generation to another during sexual reproduction. It is classic genetics that we learn at school.

In medicine, there are several thousand illnesses, often of a very low incidence (known as orphan diseases) which have a Mendelian transmission. In the most typical case there is a perfect correlation between the phenotype and the genotype: people with the mutation (or morbid allele) always have the disease and those with the disease are always carriers of the mutation.

---

1 The genotype is defined here by the sum of all the genetic information while the phenotype is the sum of all traits of a given organism.

2 An allele is a molecular form of a DNA sequence (a gene for example). A sequence is said to be polymorphic if there are several possible alleles of that sequence in a population. The allele is then one of the forms taken by a polymorphism. An allele corresponds physically to a more or less greater substitution, deletion or insertion of nucleotide bases relative to the sequence considered as the reference.

To mention just two classic and caricatural examples, mutations in the CFTR gene (coding for a chloride ion-transmembrane channel) are associated with cystic fibrosis and a specific mutation of the β chain of hemoglobin is associated with sickle cell anemia.

Nevertheless, most hereditary characteristics do not follow a simple genotype/phenotype correlation. Distortions in the relationship between genotype and phenotype are common and have been recognized for a long time (including by Mendel[3]) thus some people with mutations may not have the expected disease. Phenotype modulator factors are then used to explain these discrepancies. These additional variables make it possible to reestablish an explicit genotype/phenotype relationship. Some phenotype modulating factors are genetic. This is the case of the foetal haemoglobin gene in sickle cell anemia. The genetically programmed persistence of foetal haemoglobin limits the severity of the disease.

Environmental factors may also contribute to the phenotype. This is the case of bronchial infection with *Pseudomonas aeruginosa* in cystic fibrosis, which precipitates the progression of pulmonary disease. Some pathologies may even depend entirely on exposure to an environmental risk factor that must be added to the genetic risk for the disease to be expressed. This is the case of phenylketonuria where exposure to a high-protein diet is necessary for clinical signs to appear in mutated subjects. For celiac disease, wheat consumption by genetically predetermined people is required to cause disease. In the latter two examples, exclusion diets (phenylalanine for phenylketonuria or gluten in wheat for celiac disease) are in fact very effective in controlling the expression of the disease in genetically predisposed individuals. In all these situations, the link between the genotype and the phenotype is no longer direct and implies taking into account additional risk factors in relation to the morbid genotype.

By going further into complexity, the phenotype can be the combination of a very large number of genes and many environmental factors. This is the case for diseases known as complex genetic disorders. These diseases are common in human populations. These include diabetes mellitus, cardiovascular diseases, psychiatric disorders, autoimmune diseases, degenerative diseases, cancer, etc. To avoid multiplying examples, we will often take Crohn's disease, an inflammatory bowel disease, as an illustration. Recent genome exploration work for these pathologies shows that predisposing genes are numerous. At least several tens or hundreds for each disease, at least 140 for Crohn's disease.[4]

---

[3] Lambert, G. La légende des gènes.

[4] Jostins, L. et al. 2012. Nature, 491: 119–24.

Thus, in total, several thousand polymorphisms have been associated with complex diseases.[5] However, each of the susceptibility alleles has a weak individual effect. Thus, most genetic polymorphisms associated with complex human diseases carry an estimated relative risk between 1 and 1.5. In other words, the likelihood of developing the disease for a person with a risky genetic variant is no more than 1.5 times higher than that of a person not carrying the variant. In comparison, the same relative risk for a typical Mendelian disease is theoretically infinite. The disease appears to result from a multitude of low-risk alleles scattered throughout the genome.

The genetics of complex diseases then leads from an exact deterministic genetic model to a fuzzy probabilistic model. Instead of predicting a disease every time from simple genetic information, the geneticist must now take into account a very large amount of information, not only genetic but also environmental, to arrive at an approximation of a risk. Complex diseases require us to change the way we think about the relationship between genotype and phenotype.

Geneticists, however, believe that the link between the genotype and the phenotype is not probabilistic in nature. It is our incomplete knowledge of the genetic determinism of diseases that explains why we have to use probabilities to approach reality. Accordingly, there are still unknown risk factors which, once discovered, will lead to a complete knowledge of the determinism of complex diseases, making the probabilities of the model disappear.

In fact, we have not explored all the variations present on the genome. The polymorphisms studied to date are mainly polymorphisms with two simple alleles and common in the general population. Studies on more complex common gene variants (copy-number variations) are less numerous but produce comparable results: a large number of genetic polymorphisms with a low effect.[6] There remains the question of the rare genetic variants not yet systematically explored to date in most complex diseases.

Rare genetic variants could, if they proved to have a strong phenotypic effect, allow us to return to determinism closer to the Mendelian model and thus more explicit.[7] The current question is therefore whether rare mutations with high penetrance and which can therefore have a strong predictive effect

---

[5] For an update: http://www.genome.gov/gwastudies. See too: Visscher, P.M. et al. 2012. Am. J. Hum. Genet., 90: 7–24.

[6] The Wellcome Trust Case Control Consortium. 2010. Nature, 464: 713–720.

[7] Note that it has been proposed that the loose associations identified during the screenings of the genome with common genetic variants could be seen as the synthesis of multiple associations with several rare variants of the same gene each having a strong effect. We talk about synthetic association. For Crohn's disease, however, this hypothesis is not confirmed by the facts (see, for example, Momozawa, Y. et al. 2011. Nat. Genet., 43: 43–7).

on the disease remain to be discovered. Many laboratories are currently searching for rare genetic variations left out of early pan-genomic studies.[8]

There is no doubt that such mutations will be discovered, at least for some extreme morbid phenotypes. Thus, for Crohn's disease, in infants, it is possible to identify mutations with a strong effect, such as those of the interleukin-10 receptor.[9] Such discoveries may nevertheless remain confined to a few rare cases and not be generalizable to the majority of patients. Although the results of large-scale sequencing of the genomes of patients are not yet known, I am, in fact, rather pessimistic about our ability to discover rare and highly penetrant genetic variations accounting for a significant part of disease risk for two reasons.

My first argument is that the analysis of the genealogical trees of the majority of the patients shows that the transmission of the disease is far from the proportions defined by Mendel. Indeed, according to a model of a rare mutation with a strong phenotypic effect, even in the presence of a strong genetic heterogeneity (in the extreme, a specific mutation per family of patients) and a large number of neo-mutations, we expect to see Mendelian segregation of the character in families. We should therefore find values close to those defined by Mendel, which is not the case. This also means, in practice, that some of these genes should certainly have been identified for complex diseases by genetic linkage approaches which are very effective in identifying genes with a strong effect. However, despite major efforts, few susceptibility genes have been discovered by positional cloning for complex diseases. For Crohn's disease, for example, the NOD2 gene is a very isolated[10] example. One can always invoke other unknown risk factors in addition to the Mendelian variant studied but then take a step backwards, in disfavour of the hypothesis of highly penetrant genetic variants.

My second argument is that if there are high-risk alleles that have a strong effect, there is no reason to think that they are all rare. At least some of them should be frequent and should therefore have already been identified, which is not the case. In the case of serious illnesses, it can be argued that alleles with strong effect are rapidly eliminated from the population (and therefore rare) because the diseased subjects, bearing these alleles, die without any offspring. However, for most non-morbid (and therefore non-fatal) traits, only low-effect polymorphisms are observed. For example, height is a universal phenotypic trait with a strong genetic component. It is thus possible to predict the height of a child from the height of its two parents in a fairly reliable manner by a simple calculation.[11] Pan-genomic

[8] For a recent review see for example Wagner, M.J. 2013. Pharmacogenomics, 14: 413–24.

[9] Clocker, E.O. et al. 2009. N. Engl. J. Med., 361: 2033–2045.

[10] Hugot, J.P. et al. 2001. Nature, 411: 599–603.

[11] The formula used by pediatricians is: Child height = (Mother height + Father height)/ 2 +/−7 cm depending on whether the child is a boy or a girl.

association studies have identified at least 180 genes playing a role in genetic predisposition to height.[12] But none of these polymorphisms have a strong effect. Yet height is not a serious disease that can drastically reduce progeny as is the case with severe diseases. Strong effect polymorphisms affecting stature should therefore be transmitted from one generation to the next and accumulate over generations. We should then identify at least some of them that have become frequent by simple genetic drift or by the effect of natural selection, if we consider that height is a character giving a selective advantage (in one direction or another).

In other words, accepting the idea of strong-acting polymorphisms implies accepting the idea that some of them must be frequent. It will be noted, to reinforce this point of view, that the genetic variations associated with the complex phenotypes currently identified indeed take all possible frequencies, from rare to very frequent.[13] According to this reasoning, some high-effect polymorphisms should already have been discovered in view of the very large number of studies available for a very large number of complex traits. Unless otherwise noted, this is not the case, suggesting that the genetics of complex traits will not be solved by the discovery of new alleles with a strong effect for most patients.

Whatever the case, studies in progress will soon be decisive.[14] In the meantime, let us continue to advance with our current knowledge by considering that we will not return to a more or less simple Mendelian model for complex genetic diseases.

Another way of reducing the current probabilistic model to a more deterministic model would be to identify particular combinations of genetic and/or environmental factors associated with very high risks of getting sick. We have given some examples of such combinations for celiac disease or phenylketonuria. Unfortunately, such combinations are difficult to identify for complex genetic disorders and it has generally not been possible to determine recurrent allelic combinations in patients. Current methods lack power, however, and new mathematical approaches may challenge this temporary conclusion. Geneticists therefore hope to show that the univocal genotype/phenotype relationship as described by Mendel can be replaced by a relation "combinations-of-multiple-alleles-and-risk-factors-environmental"/univocal phenotype.

However, it is not very realistic to think back to a strict deterministic model and it is very likely that the links between allelic combinations and the phenotype will remain largely probabilistic. Thus, for complex diseases and unlike simple Mendelian diseases, one must resolve to be able to

[12] For a recent review see Letter G. Hum. 2011. Genet., 129: 465–472.

[13] Jostins, L. et al. 2012. Nature, 491: 119–24.

[14] Initial results suggest that rare alleles have a limited effect on the phenotype at the population level: Morrison, A.C. et al. 2013. Nature Genet., 45: 899–901.

propose only probabilities to predict a morbid phenotype when one knows the genotype of a given individual.

As mentioned above, for the "height" character, there is a very large number of susceptibility genes, each with a low phenotypic effect. The same is true for corpulence or for complex genetic diseases. One can then wonder about the absolute number of genes predisposing to complex phenotypes. To answer this question, it must be understood that the definition of a "susceptibility gene" is only statistical. Thus, for Crohn's disease, there are now just over one hundred genes at risk, but if we did the same studies with tens or hundreds of thousands of additional patients, there is no doubt that we increase this number simply by increasing the power of the statistical test used. Indeed, we find other genes, not yet accepted as "at risk" because they are more difficult to validate due to an even lower effect on the phenotype. Alternatively and more simply, it would be enough to change the statistical threshold retained by the researchers to define what is an "at risk" allele, to modify the number of recognized susceptibility genes. In fact, and contrary to the Mendelian characters, there is no clear limit between a gene at risk and a gene not at risk. It is only a matter of defining at some point in medical research. But then, could not all genes be considered at risk for Crohn's disease?

For height, some authors have thus posed the question of taking into account not only the polymorphisms "validated" statistically as "at risk" but on the contrary the greater part of the information available on the whole genome. They have shown that the hereditary predisposition explained can be significantly increased if one takes into account the set of all the common polymorphisms of the genome.[15] This idea suggests that a complex phenotype is due to a very large number of alleles and not to a small number and that it is illusory to imagine placing a line of demarcation between an "at risk" gene and a "not at risk". In other words, if one observes that no polymorphism has a strong effect, one can also say that no polymorphism is strictly neutral. But then we must go so far as to think that all the genes of the genome have a more or less noticeable effect on the phenotype studied. We are far from the Mendelian model with unique or few polymorphisms contributing strongly to the variance of the character!

Another observation supports the idea that the genes participating in a phenotype are very numerous. Human chromosomal abnormalities are accompanied by symptoms that most often associate: (i) organ malformations; (ii) stunted growth retardation; (iii) retarded development of psychomotor functions; and (iv) a disharmonious face. The most common and known chromosomal abnormality is trisomy 21, which illustrates this fact perfectly.

[15] Yang et al. 2010. Nature Genetics, 42: 565–9.

However, databases of medical genetics confirm that this case is not an exception but rather the rule.[16] From this observation, it can be said that any (somewhat large) region of the genome contains one or more sequences involved in organ formation, psycho-motor development and height. Here again, the conclusion is that it is illusory to think of a complex phenotype as borne by a small number of genetic variations if most regions of the genome can contribute to it.[17]

In the end, it seems that we must accept the idea that multiple genes (possibly all the genes of the genome) participate more or less in the definition of a phenotypic character. By immediate deduction, it is then necessary to imagine that all genes participate more or less in all phenotypic characters. Any genetic information contributes to the definition of any phenotypic trait! This degraded causality is in practice the opposite of the classic Mendelian model where a single gene defines a single phenotype. On the other hand, it is compatible with the modern vision of biological network organization in living organisms where most genes or gene products are interconnected and thus contribute more or less to all biological functions. We will come back. It is also consistent with the observation of a probabilistic genotype/phenotype link suggested by the genetic data available today.

In the absence of genes having a strong predictive power on the phenotype, strong environmental determinism can be invoked for complex genetic diseases. There is no doubt that the environment plays an important role in a variety of diseases. The environment also plays an important role in the development of non-morbid traits. We have cited the example of the gluten required for the development of celiac disease, but other examples are well known, such as tobacco for lung cancer, saturated fat for atheromatous disease, nutritional status for height, modern western lifestyle for Crohn's disease, etc. These elements of the environment are important and it is essential to make progress on their identification.

However, most known environmental risk factors also have weak effects and the above discussion of the results of genetic studies is largely applicable to epidemiological data. Moreover, additional environmental risk factors will probably not be sufficient to completely solve the problem of determinism of complex genetic traits. The latter are usually hereditary and must therefore be determined by heritable factors, which is not often the case for environmental factors. There is now a difficult problem for geneticists, known as missing heritability, which corresponds to a lack of knowledge of what contributes to the hereditary part of complex diseases

[16] http://decipher.sanger.ac.uk/.

[17] Recent studies confirm that the phenotypic variance of complex traits is proportional to the size of the chromosomes. The genetic variations involved in complex diseases are therefore well dispersed throughout the genome.

and traits. And it is difficult to imagine filling the gap with environmental factors shared between affected relatives.[18]

The problem of missing heritability can be summarized as follows. The discovery of polymorphisms associated with susceptibility to complex human genetic diseases revealed that the heritable variance of the morbid character attributable to these numerous genetic polymorphisms is much lower than the whole heritability expected on the basis of theoretical calculations.[19] In fact, two relatives are more alike than expected at the phenotypic level, given their measured genetic resemblance. In other words, the resemblance between relatives is poorly explained by the transmission of the only known genetic polymorphisms. There must therefore be an additional cause of resemblance between the relatives that is not identified. This question has been much debated for morbid phenotypes, but it is also true for non-morbid characters like height. Thus fewer than 20 to 30% of the expected heritability is explained by the genetic polymorphisms currently identified for height. Often even, only a few percent of heritability are explained as in the case of corpulence. We have seen for the height that taking into account a very large number of genetic polymorphisms (or all) makes it possible to increase the proportion of explained heritability. However, this does not seem to be already sufficient.

Since heritability is supposed to reflect the "inheritable" part of the variance of a phenotypic trait, it is necessary to explain the missing heritability by hereditary factors.

Some environmental factors may be considered inheritable (a table set, for example!). As mentioned above, however, it is difficult to imagine that most of the heritability is not carried by the egg itself but is carried by environmental factors characteristic of a family and transmitted from generation to generation. Consequently, it is difficult to believe that the essential part of the hidden inheritance of the phenotype is to be sought in the environment. This long reasoning shows that the genotype and the environment do not explain the transmission of complex genetic traits on their own and that this point requires an explanation.

As we have seen, in the present state of genetic and epidemiological knowledge, we are obliged to note that, except for simple Mendelian characters, the genotype/phenotype relationship can be defined today only in a probabilistic way. From a pragmatic point of view, this justifies the statistical approach currently being developed by geneticists who study a very large number of people (sometimes several tens or hundreds

[18] See for example Maher B. 2008. Nature, 456: 18–21.

[19] It has been suggested that the problem of missing heritability is a false problem and that heritability calculated on family datasets is overestimated. However, alternative methods of heritability computation only partially solve the problem: Zaitlen, N. et al. 2014. Nature Genet., 46: 1356–62.

of thousands) in order to define more precisely the empirical distribution of risks in a population. This statistical approach, however, does not have the ability to predict the development of a phenotypic character (morbid or otherwise) on an individual scale. The calculation of the frequency of an event in a population does not allow an answer on a personal and unique situation. This question is important because it forms the basis of the concept of predictive medicine which was dear to Jean Dausset, co-discoverer of the HLA system and Nobel Prize in medicine, and which is currently widely promoted by numerous public and private research groups. Predictive medicine should aim to provide an individual prediction for healthy subjects on their own risk of becoming ill so that they eventually adapt their lifestyle based on this information.

The question is whether a given person will become sick or not aware of the sequence of his genome and its exposure to environmental factors (and in this case, how to remedy them). From an epistemological point of view, the question posed today to health researchers is thus to construct a completely deterministic model to reliably predict the occurrence of an event (the disease in our example) from genetic and environmental information. Today, however, a genotype can be considered at best only as a propensity to develop a phenotype, but it does not allow a deterministic prediction.[20] It only signifies the authorized phenotypic possibilities and their probabilities, which poses a difficult problem for doctors and geneticists (and even more so for their patients!).

The question of the probabilistic nature of a phenomenon (here the realization of a phenotypic trait) reminds us of the one that animated the debate of quantum physicists in the 20th century (in particular between Bohr and Einstein). The question for quantum physicists was whether the state of a particle is essentially probabilistic or whether the probabilistic view we have of the state of a microscopic system reflects only our ignorance of hidden variables that must be added to the model to make it completely deterministic. It is the famous question of whether God plays the dice that we could paraphrase in biology by "does life play dice?". In the present state of things, most physicists accept the essentially probabilistic character of the state of a particle. To date, most biologists would not bet on the essentially probabilistic nature of the genotype/phenotype relationship. This attitude is a position which, although philosophical, seems reasonable and follows the classical tradition of scientific research based on the principle of strict causality. The impossibility of predicting an individual phenotype is thus considered due to the fact that it is the result of a very large number of parameters, impossible to predict exactly as for a weather forecast.

[20] For a discussion of the significance of probabilities in biology see for example Martin, T. in Kupiec, J.J., Gandrillon, O., Morange, M. and Silberstein, M. Le hasard au Coeur de la cellule, Chapter 2.

In any case, whatever the reason, it is necessary to consider, at least temporarily, that the nature of the genotype/phenotype link is probabilistic. It is therefore necessary to take the view that, for a given individual, it is not possible to accurately predict its phenotype even after sequencing its genome and analyzing its entire environment. In other words, the probabilistic nature of the link between genotype and phenotype must be assumed today as a given of the problem.

In summary, our recent acquisitions on hereditary predisposition to complex genetic traits require a thorough review of the genotype/phenotype link. The probabilistic character of the relationship between the genotype and the phenotype must be taken into account as a given of the problem. It must be admitted that a large number of genes participate in most phenotypes. Finally, we must explain the apparent missing heritability of the characters.

Part 1

The Genotype/Phenotype Relationship

The problem of the relationship between the genotype and the phenotype is probably one of the most important in modern biology. The genotype/ phenotype relationship is at the heart of the issues of the formation and maintenance of living organisms, diseases, heredity, speciation and evolution of living organisms. It is a question of understanding the link between the genetic information and the self-organized biological network that make up the organism.

1

The Beginnings of Life

What is life? It is the brilliance of a firefly in the night. It is the breath of a buffalo in winter. It is the little shadow running in the grass and lost at sunset.

Isapo-Muxika

The origin of life is an exciting but difficult question to solve. It would have appeared on Earth 3.8 billion years ago. For some, life was originally carried by an exoplanet and scattered through the cosmos to the earth. This theory of panspermia is supported by the observation of amino acids and possible bacteria in meteorites. Another possibility is the gradual accumulation of reactive biological molecules in the ocean of the primitive Earth. This is the hypothesis that has been the subject of the greatest number of studies, in particular those of Miller. Eventually, life may have appeared in the depths of the earth. There are increasingly convincing arguments for this assumption. Thus, the depths of the oceanic faults formed (and still form) a relatively stable, protected environment (against ultraviolet rays in particular) and paradoxically quite welcoming for a biological life built on the metabolism of sulfur while the surface of the Earth was inhospitable in the first billion years of its existence. Ocean faults prove that they contain stable and complex ecological systems and bacteria are found at increasing depths beneath the earth. If the latter theory proves correct, it allows us to think that life may have appeared on many planets.[21]

However it appeared, life probably began with compartmentalization of the external environment to make it an "inner medium", thus defining

[21] For more details on the origins of earthly life, the reader will find some reference works in the bibliography section.

a protocell. The compartmentation that we know today is based on lipid layers that form the plasma membrane. It is possible that the first methods of compartmentalisation were different (based on mineral layers or drops) but the lipid layer became the rule in living organisms (with specificities in archaeobacteria). We can therefore imagine the primitive cell as a sac with a membrane. If the membrane is asymmetrically permeable and/or if there is a capacity for assembling the small molecules inside the protocell, then the internal environment is different from the external environment. The notion of compartmentation leads to the notion of individuation of a entity entities, endowed with a history and a differentiated inner environment that can be seen as an organism.

This configuration of a closed environment, separated from the outside by an asymmetrical membrane, is favorable to organic chemistry because it concentrates the elements, it puts in contact various substrates and then favors the metabolic reactions. It forces cooperation between molecules. From this, one can imagine that metabolic cycles were set up.[22] The protocell has, to some extent, domesticated organic chemistry.

If life comes from Earth, its relative rapidity of appearance after its birth (less than 700 million years!) testifies to the very high efficiency of these protocells in organizing chemical reactions and evolving towards living organisms. The fact that the protocell has been able to organize a network of metabolic reactions in such a short time seems at first surprising. It must be understood, however, that the forced union of organic molecules in the protocell favors the creation of new molecules at an exponential rate. To understand this issue, it is possible to use computer modelling of networks (here of molecules). We then observe that the progressive addition of connections in a model network leads inexorably to the constitution of a self-organized system that emerges suddenly for a critical value of connectivity.[23] The relatively rapid emergence of a metabolic self-organization may thus not be as unexpected as one might think. The condition *sine qua non* is, however, a partnership between different organic molecules forming molecular networks. Life is necessarily the property of a complex system made up of heterogeneous elements united by functional interactions, in other words, a networked system.

As with any object delineated in interaction with its environment, there must have been incoming elements (small inorganic molecules, energy, ions, water, etc.) and elements accumulated and/or re-emitted in the external environment. The system must be open from a physico-chemical and thermodynamic point of view. Therefore, it seems realistic to think

[22] For an approach on the evolution of cellular metabolism see for example Cunchillos C. Les axes majeurs de l'évolution du métabolisme cellulaire in Tort P. Pour Darwin.

[23] See Gribbin, J. Le chaos, la complexité et l'émergence de la vie, p. 236 and following.

that the cell had to store its own organic production, the biggest molecules probably being the most suitable to be sequestered in the protocellular bag or exported to its contact. In our time, an identical behavior is found in plants where the complex carbohydrates (cellulose or lignin) can be seen as waste of photosynthesis that the plant stores until it is closed and prohibits mobility and neuro-motor evolution.

For a protocell, waste from organic chemistry was probably the opportunity to build the complex macromolecules we know, of which the two main forms are proteins and nucleic acids. It should be noted that certain nucleotides (ATP and GTP) which constitute DNA are universal energy reagents of living organisms. They are used to exchange energy in many cellular chemical reactions. One can then think that the DNA, macromolecule often considered as the noblest of our organism could be, at the outset, only a solution of storage. The lipids were used in the manufacture of membranes and in the production of energy. Carbohydrates were used primarily to store and use energy. The intrinsic properties of the macromolecules were thus exploited by the protocell, just as a more evolved organism exploits nature according to the properties of the objects sought.

Living organisms have transformed the physico-chemical properties of organic molecules into biological functions. This is the essential genius of the living. It is both the result and the cause of the association between multiple elements, each of which assumes an integrated functional role on a higher scale.

It is more particularly the invention of nucleotide and peptide chains which enabled the emergence, the continuation and progressively the improvement of functions that have become the support of current life (we will not therefore speak of other macromolecules such as complex carbohydrates for example). Proteins are very diverse in form, size and functional properties. They are at the basis of all the major cellular functions. They organize metabolic cycles through enzymatic catalysis. They form the internal structure of the cell with molecules of structure. Finally, proteins regulate cell exchanges with the outside and transmit information within the cell. One sees that the proteins have above all an operational role. DNA, an inert and passive molecule, has taken on an informational role without its own operational role. The cell functions were thus separated in the cell between the two large groups of macromolecules. It is because it has both operational and informational status that RNA has been proposed as a key molecule at the very beginning of life.

It is difficult to define life other than by the common properties observed in all living organisms. The living being is defined by the presence of all these properties in the same object. The most commonly shared properties among living organisms are as follows:

1) birth from another living organism. There is continuity of the living without spontaneous generation. It is not possible to construct *de novo* and by chance a modern living organism. This implies that the organization of the living being is transmitted and that there exists a filiation that unites the beings. Each organism is always a link in a lineage. It is a historical object and its organization is based on information acquired from another organization;
2) the stability of its organization. The organism is capable of variations over time by growth and/or transformation. The organism therefore has an internal plasticity. However, this plasticity usually takes place within relatively defined limits, and it can be said that living organism ensures from one moment to the next the stability of its structure. This stability implies the sustainability of an organization that can be considered as a referent. It also implies an adjustment of its internal functions to changes in external and internal conditions;
3) nutrition by assimilation of exogenous materials. This characteristic shows that the maintenance of the stability of the living being is an active phenomenon. The body constantly needs to keep itself stable from using energy and materials coming from the outside environment. Its self-organization is indeed thermodynamically far from equilibrium. Organizations surf on the wave of life, each according to a precarious balance. This point is partly the corollary of point 2;
4) multiplication by progeny giving rise to other living organisms, most often similar. This point is the corollary of point 1; and
5) inescapable death by the complete destruction of internal structures. This point is the corollary of point 3.

One can then summarize the living being to an organized object, intrinsically unstable but capable of actively maintaining its structure throughout its life and through reproduction.

One may wonder if the protocell was fully alive. It is highly probable that the networking of the macromolecules in the protocell must have provided sufficient stability to last for a certain time. It would have been, however, much less stable than a modern cell and the DNA molecule certainly contributed greatly to cellular stability.

It is probable that the evolution of the protocells was towards a gradual separation from the thermodynamic state of equilibrium. Thus, the first cells could not be far from this state of equilibrium. Progressively, the development of increasingly sophisticated metabolic cycles and the asymmetries of permeability of the plasma membrane (through transport proteins) would have moved the protocells away from this state of equilibrium. The protocell then would have needed an increasing amount of energy to keep its cycles away from equilibrium. The evolution of life can then be seen as an increasingly reliable stabilization of a self-organized system increasingly distant from the state of thermodynamic equilibrium.

A cell can be understood as a network of molecules interacting with one another. Some molecules are used for support, others for energy storage, compartmentalisation, etc. The stability of the network is brought about by the number of different molecules, the richness of the functional links that unite them, the redundancy of the possible paths connecting two molecules, the presence of feedback loops, etc. The DNA molecule probably contributed to this functional network only belatedly, when the mechanisms of regulation of the transcription were put in place. Genetic information has, however, established a reliable physical network by allowing the synthesis of identical proteins throughout the life of the individual. Without this stability, the network would have required significant and incessant readjustments. By analogy with other networks, one can imagine an air network where the airports would all be in perpetual motion, able to accommodate a few flights one day and the next day thousands. Such a network could work but would be unstable in nature, apart from any necessary adjustments to changes in passenger flows or meteorological incidents.

The protocellular sac served as a closed place conducive to the establishment of an intimate cooperation between various molecules. At the beginning, this cooperation revealed an organized structure that we call the cell and which has emergent properties that do not have molecules taken in isolation. From the outset, life is a matter of co-operation (this is true not only of the living but also of the matter which rests on the cooperation between atomic or subatomic particles). This co-operation is itself due to the constraint of sharing space and resources. At all levels, from the macromolecule to the ecosystem, it is this constraint that obliges living organisms to cooperate and to integrate functionally, bringing out at each hierarchical level of life a new and larger and more complex organism.

Little is known about the first modalities of molecular cooperation between proteins and nucleic acids. Nevertheless, a chemical symbiosis has developed between the two major classes of macromolecules. This symbiosis was sealed by the appearance of the genetic code and other mechanisms of transcription and translation. The links between DNA and proteins then became the following: DNA stores information on the primary structure of the peptide chain while the peptide chains make (replication and transcription) and use for their own synthesis (translation) nucleotides.

It was initially a good deal for proteins that had an effective method to synthesize non-randomly and thus ensure stability to cellular functions. But the DNA molecule took on a second function that altered the balance of forces. This second function became essential when effective cell replication emerged. Initially, the primordial protocell was to divide asymmetrically once its size became too large to allow sufficient biophysical stability to the protocellular sac. Each daughter protocell then had to carry away a portion of the mother protocell.

This portion of cytoplasm and membrane was often too unbalanced for the two daughter protocells to survive. One can imagine that sometimes enzymes or essential transport molecules were missing by simple random sampling during division. The cells carrying the DNA molecule(s) then had to be advantageous since they were less dependent on these sampling biases by having the possibility of restoring the missing protein with the invention of the translation. Natural selection therefore had to appear with the DNA molecule. The proteins not listed on the DNA molecule were then extinguished and the DNA, however inert and passive, became to some extent the master of the game. Thus, if the proteins controlled the self-organization of the organism, DNA controlled the perpetuation of proteins over the lifetime and from one generation to the next. The proteins then had to replicate the DNA to keep themselves from one generation to the next.

At this almost catch 22 game, the proteins could have regained control by modifying the DNA. It is indeed they who decide on replication and its modalities. They could thus have introduced genetic variations and recreated new peptide sequences from a DNA sequence largely modified or even made *de novo* in each generation. The disadvantage of this process is that the function of a protein is not predicted by its primary structure. However, introducing random changes in DNA is most often deleterious and seriously destabilizes the molecular organization of the protein. The cell has therefore had a greater advantage in maintaining reliable replication rather than inducing errors except in certain special cases such as extreme stress in bacteria which induces major molecular lesions where replication errors may appear as a last chance rescue.

This division of labor became a rule that was declared "central dogma of molecular biology" by Francis Crick in 1957.[24] Simply stated, he says the gene is the cause of the protein. By extension, since all proteins depend on genes and all biological functions depend to one degree or another on proteins, the central dogma of biology has become generalized: "the genotype is the cause of the phenotype". We see that the central dogma of biology is asymmetrical. The protein, although it serves to replicate the DNA molecule, does not carry information to make the gene. It is therefore considered a product.

The central dogma of biology is in fact centered on hereditary information. The theory of evolution has such an important place in biology that biologists usually consider that reproductive success is what

[24] "DNA makes RNA; RNA produces proteins. And proteins make us" see Fox Keller, E., the century of the gene, p. 55 of the French edition.

characterizes the living being. This explains why the generalization of Francis Crick's dogma was naturally accepted. However, the primary purpose of life is not to reproduce itself, but first to maintain oneself, that is, to self-fulfill, then, if possible, to multiply. In this case, according to a vision centered on the operational aspect of the cell (and therefore the protein), the phenotype is no longer a product but an actor that uses the genotype as a tool. It is the protein that replicates and uses genetic information. From this point of view, the encoding function allowing the translation of DNA into protein becomes more important for the living being than its replication function allowing the reproduction from one generation to the next. There is, in fact, much work to demonstrate that the cell regulates very finely the use of the genome by various epigenetic processes. So we should not be too dogmatic!

In fact, the cell specialized the categories of macromolecules to perform functions that were necessary to it. Proteins realize the phenotype, the nucleic acids preserve the information necessary for self-realization. A bit like a RAM and a ROM in a computer. However, it does not make sense to maintain information if it is not used for the realization of a phenotype. Conversely, there is no organization without information allowing self-realization. Each has its role and it is because the cell imposed a functional specialization on its macromolecules that they were condemned to a chemical symbiosis where each partner is indispensable to the other.

A large number of biological issues is a question of the duality between the information and operational visions of living organisms and raises the question of the duality between the genotype (or genetic information) and the phenotype (or the self-organized molecular network).[25] This is the issue we will address throughout this text.

[25] This issue is discussed in more detail in Maynard Smith, H. and Szathmary, E. The origins of life, Chapter 1.

2

Genetic Program and Self-Organization

Successes and excesses of molecular biology have partly been built on the computer metaphor of the genetic "program".

Henri Atlan

A living organism is defined by its phenotype which is the set of observable characters in its structure and functioning. But it is clear that this phenotype depends on the genome that is transmitted by the parent(s). Dogs do not make cats! Consequently, an organism is also defined by its genome, which carries the genetic information (hereinafter referred to as the genotype in the remainder of this text). The question asked is to understand the relationship between the genotype and the phenotype: what are their links?

The currently most shared paradigm is that of the genetic program.[26] This expression seems to have been used for the first time, a little over 50 years ago (in 1961), independently by Mayr and by Jacob and Monod.[27] According to Jacob, the organization is "the realization of a program prescribed by heredity" (genetically understood). For Mayr, it is "coded or ordered information that controls a process (or behavior) toward a goal. The program not only contains the molding of this purpose, but also instructions

[26] Several publications address this issue in depth. See for example Longo, G. and Tendero, P.E. in Miquel, P.A. "Biologie du 21ème siècle. Evolution des concepts fondateurs", p. 185.

[27] Fox Keller, E. The century of the gene. French edition, p. 79.

on how to use this molding. A program is not a description of a situation but a set of instructions."[28]

The idea of a genetic program has been accepted very willingly by the scientific community because it merely generalizes the central dogma of molecular biology to complex organisms: the genotype explains the phenotype. The notion of a genetic program means that, from a genome, it is possible to build a living organism from scratch. All instructions are contained in the DNA. In other words, the organism, in its structure and function, results from the deployment in time and space of coordinated and precise information carried by the genes. Genetic information is explicitly or implicitly compared to a computer software that deploys to produce a digital work. Moreover, the simultaneous developments of molecular biology and computer science have favored the analogical reasoning between the genome and the computer program. Ultimately, the idea of a genetic program made it possible to respond to the teleological question of the creation of living organisms in a "scientific" and cheap way (i.e., without using a superior entity): the genome is necessary and sufficient to create a living being.

Monozygotic twins or true twins are the example from nature that best illustrates the idea of genetic program. True twins are genetically identical. They are also phenotypically identical throughout their lives and even if they are bred separately. Certainly there are some differences between them (for fingerprints or complex diseases for example). However, these are points that can be described as minor. The very great resemblance between monozygotic twins argues in favor of an extremely strong genetic determinism in the realization of the living. The systematic resemblance between organisms of the same species (i.e., sharing the same gene pool) is also a very important argument in favor of a strong genetic determinism.

The experimental results obtained over the past fifty years have reinforced the paradigm of the genetic program. The discovery of the homeotic genes that direct the major stages of embryogenesis has established that genes are essential for ontogenesis. They were thus assimilated to the lines of code of the genetic program.[29] Advances in molecular biology and then in synthetic biology have shown that it is possible to modify the properties of living organisms by modifying its DNA. Finally, cloning experiments validated that the introduction of a differentiated cell genome into a fertilized egg results in reproduction of the cloned individual. Thus, all these scientific observations strongly argued in favor of the hypothesis that the DNA determines the phenotype precisely and reliably.

[28] Mayr, E. After Darwin. French edition, p. 50.

[29] See for example Chaline, J. and Marchand, D. Les merveilles de l'évolution.

However, the concept of genetic program has been criticized.[30] Initially for the very unlikely character of building a DNA molecule that would have a chance of realizing a coherent living organism if one would randomly draw the nucleotides to form this long molecular chain. This improbability results from the large number of nucleotides forming a complex genome (3 billion nucleotides per haploid genome for humans). Armies of monkeys were then called to write random lines of letters to explain that the living could not arise by chance. The conclusion was that the genetic program had to have a purpose and that it was therefore necessary to call upon a higher creative entity to explain the living organism as well as to use programmers to explain a computer program. However, if the living being is clearly not the product of immediate chance, this does not mean that there is no way to achieve a complex genome other than through the existence of a higher entity. On the contrary, all the work of biologists on evolution shows that the latter explains the appearance of a progressively increasing complexity of the living. Computer simulation experiments confirm that evolution can lead to increasing network complexity without the use of superior intelligence. It is therefore very likely that the DNA molecule needed to produce complex living beings has been created over time by evolving mechanisms not preconceived.

It has also been argued that the genome alone cannot build a living organism. It must be in a cell that can interpret it. This is a reminder of the indissociable nature of the informational and operational aspects of life. Here, too, this argument does not call into question the analogy with the computer program. Just as a genome needs a cell around it, software needs a computer to be exploited and result in a digital work. The question of knowing the precise role of the cell itself in the expression of its genome and therefore in the initial implementation of the genetic program is important, but it does not fundamentally call into question the notion of the program.

The strongest arguments against the paradigm of the computer program are in reality of a different nature. They question the idea of a predetermination of life itself. They therefore run counter to the concept of information capable of predictably organizing a living organism. The argument does not focus so much on the genotype's information capacity to explain the living. It is indeed universally accepted that hereditary information is necessary for the formation and perpetuation of a living being. It is rather on the notion of program that the question stumbles. The program involves a definite and repeatable succession of actions prefigured in advance. Thus, any program idea whether genetic, essential or divine is refuted, considering that the living cannot be predefined and therefore programmed.

[30] Miquel, P.A. in "Biologie du 21ème siècle. Evolution des concepts fondateurs". Chapters 7 and 8.

This opposition to the concept of program is supported by multiple observations. Thus, the plasticity of living beings implies that the genotype is not the only parameter that participates in the definition of the phenotype. The examples of phenotypic plasticity (or polyphenism) are numerous and this subject has been perfectly detailed by West-Eberhard.[31] By way of example, for the same organism, plasticity is exercised throughout life in the stages of development (embryo, child, adult, elderly); environmental variations (muscle mass or adipose tissue); physiological (puberty, pregnancy, menopause) or pathological (disease, aging) events. It is responsible for the asymmetries of the body (asymmetry right-left of a face for example), transformations of the insects (caterpillar-cocoon-butterfly), the formation of organs in the plant, etc. Plasticity is responsible for sexual dimorphism (especially when it is determined by the environment as in crocodiles), the dimorphism of social insects (queen, warrior, worker). These multiple phenotypes observed for identical (or quasi-identical) genotypes indicate that there are other factors involved in the phenotype. Genetic information alone is therefore not sufficient to explain the phenotype. A genetic program based on the environment must be used.

Moreover, and above all, another argument is against rigid programming. There are exogenous and/or stochastic events at any time that call into question the precision of the program. The impact of environmental variations is well demonstrated at all scales of life. For example, protein conformation depends on pH or ionic concentrations. Stochastic fluctuations have also been demonstrated. For example, the expression of the genes by cells in culture (and therefore under perfectly identical conditions) is variable from one cell to another. The so-called illegitimate expression of genes in tissues where they should not be expressed also testifies to a certain indeterminacy. This indeterminacy is due to multiple causes.[32] We can mention the small number of molecules (80% of the proteins of a bacterium are present at less than 100 specimens); molecular bulk in the cell; the non-optimal nature of molecular interactions; etc. It is then difficult to understand how a succession of predefined information can control the realization and maintenance of a living organism on its own. Here we find the same problems as in meteorology where long-term prediction is impossible because of the incalculable consequences of modest local variations.

This difficulty in predicting comes largely from the complex nature of living systems. Genetic information is a succession of symbols, which, transcribed and/or translated by the genetic code, give a nucleotide and/or peptide sequence. But beyond this, the conformation of proteins or RNAs

---

[31] West-Eberhard, M.J. Developmental plasticity and evolution, Chapter 3.

[32] For a more detailed discussion see Heams, T. in Kupiec, J.J., Gandrillon, O., Morange, M. and Silberstein, M. Le hasard au coeur de la cellule.

with enzymatic activity (such as ribozymes) or even the effect of micro-RNAs on gene expression is not immediately deducible from their sequence. The function of the elements coded even less. The cell morphology derives from all the products coded even less. The organization of the cells among them even less, etc. At each hierarchical level of the life-scale, from the molecule to the animal society, it becomes increasingly difficult to understand the deterministic link between the genotype and the phenotype. Causality degrades. Yet a termite mound can easily be recognized when walking in nature. But what biological information can well encode a termite mound?

Similarly, if causality degrades, how can the phenotypic resemblance between relatives and organisms of the same species be explained? For this, some propose to call upon archetypes or an essential nature of the living which would serve as a kind of mold. But then the problem of the genetic program is moved back to the notion of Platonic archetypes, thus returning to another form of preconception of the organism.

If a program cannot maintain a sufficient degree of accuracy, it becomes necessary to resort to a self-realization of the living being that allows it to be built or maintained by itself, using methods of regulation. Researchers then worked on the hypothesis that the phenomena of spontaneous self-organization can account for the living. The phenomena of self-organization can be very precise, as evidenced by the regulation of physicochemical parameters of complex biological ecosystems and even of planet Earth as a whole (see the Gaia hypothesis).[33]

In order to account for stable and reproducible figures, systems of automata such as Boolean automata developed by Stuart Kauffman[34] have been used. These automata are made up of individual elements called cells that interact with each other according to simple rules.[35] One can then follow the evolution of such a system by letting it react to the rules programmed on the computer. Some rules lead to stable multicellular structures such as cellular monolayers, for example. These structures are predictable (we speak of attractors) if we know the structure of the initial network and the rules of interaction between cells. It is therefore possible to make complex structures emerge from simple and few rules of self-organization.

These models of spontaneous self-organization are important but there is still far to go before accepting them as valid models for the living. Thus,

[33] See James Lovelock. The Gaia hypothesis. It should be noted, however, that the planet Earth does not reproduce itself in the same way as a genealogy. It also does not maintain its structure unchanged throughout its existence, as evidenced by the current global warming. It is therefore not shown that the results of his self-organization are predictable as for a living being.

[34] Lambert, D. and Rezsöhazy, R. Comment les pattes viennent au serpent, p. 223. Gribbin J. Le chaos, la complexité et l'émergence de la vie, p. 246.

[35] An example of a rule could be the following: «a cell located between two cells with a common phenotype must adopt this phenotype while if it is between cells with different phenotypes, it remains unchanged».

the very large number of molecular and cellular types implies complex rules from which it becomes difficult to predict evolution. Above all, for there to be some permanence and heredity of the complex system, it is necessary to admit in principle the permanence and a transmission of the self-organization in the form of parameters defining the network and rules of interactions between cells. Besides, the experiments of self-organization are realized *in silico*, on computers, using programs. We then find the idea of the heredity of information allowing the realization of the living.

It must therefore be admitted that self-organization and hereditary information are necessary to explain the living. This duality is not only necessary but is observed and verified through all the facts discussed above. But then how to reconcile the points of view? Basically, if there is no possible prediction from a genome, it means that it has to be experimented, i.e., put it in a situation of becoming a phenotype to know its nature. A genome must be put into an egg so that it is realized (in the strict sense of reality). There is then necessarily self-realization whenever a living being is born and develops from its genome. To take up the idea of genetic program, a living being is like a console game hero who would hold the controllers himself!

The problem is then to find a conceptual point of view that solves the dilemma of the existence of both an informational determinism and a spontaneous self-organization, more or less free and adapted to varying environmental parameters.

3

The Phenotype

The donkey is as sensitive, as patient, as calm as the horse is proud, ardent, impetuous.

Buffon

The phenotype can be seen as everything that constitutes the structure and functioning of a living organism. It is therefore by definition variable from one moment to the other of the life of the organism. The phenotype is the result, at a given moment, of the entire pathway experienced by the organism. It can also be described as the momentary state of the self-structured network of molecules and cells forming the organism. Finally, it can be defined by the measurement of physico-chemical parameters characterizing the organism at a given instant.

The character or trait defines a unit of phenotype. Like the gene, character is a fuzzy concept, which gives it a strong operational power. Thus, the classification of species relies essentially on the notion of characters. Certain traits may be common to taxa of different levels (kingdoms, phyla, classes, orders, families, genera or only species). For example, an animal is a mammal if it has hair. Other traits are specific to a species such as the song of a bird or only a variety like the color of a plant. There are relatively simple phenotypic characters such as the ability of bacteria to grow in a given medium but most often the characters are complex (especially in eukaryotes) such as bone shape, physiological parameters, diseases, etc.

This complexity can be partially approximated by reducing the complexity into simpler and supposedly more fundamental elements (commonly called intermediate phenotypes in genetics) which we will call elementary characters here. Elementary characters are often difficult to define in practice, but for example, several anthropometric measurements can be used to assess the size or shape of a living structure, a morbid trait can be divided into functional modules, the concentration of several metabolites may explore the activation of a metabolic pathway, etc. Finally, the global phenotype of the organism can be described at any time by a set of values observed on N elementary traits whose number is large but in principle countable.

The living is built on a hierarchy of organized biological units ranging from the macromolecule to the animal society and ecosystems. At each level of integration corresponds a system composed of lower-level units which form a whole with emergent properties (macromolecule, cell, tissue, organ, individual, species, etc.). The most basic units in the hierarchy are naturally the oldest and least complex. It is not necessary here to discuss the number of hierarchical levels or their precise definition.[36] The hierarchical organization of life from biological units of increasing complexity is a very generally accepted fact.

As we have seen, living organisms are complex systems in which it is increasingly common to reason in networks. This way of thinking is, of course, linked to the development of networks in the economic, social, computer and other sectors. It is therefore natural that biology should also be conceived with the same conceptual tools at the beginning of this century. Biological networks are multiple. They can be metabolic networks, supporting the small molecules that form the bricks of the living. These may be structural networks that form the architecture and form of the organism. They can be signalling networks that carry information in the cell and between cells. These may include social networks for animals sharing the same territory, etc. These biological networks consist of nodes representing the living elements (proteins, cells, organs, organisms, species) and of links (interactions between nodes signifying a reaction or a physical association for example) through which flows of molecules, information, mechanical strength, etc. One of the advantages of the concept of networking is that it can be applied at multiple scales and for various biological units.

It has been shown that biological networks are "scale-free".[37] This means that the relationship between the number of nodes and their connectivity follows a power law. In other words, there are very few nodes with a very large number of connections while there are many nodes with little connection to their neighbors. The ontogenesis of such networks could be based on a simple rule that is comparable to "one only lends to the rich". It is enough that the most connected node is chosen preferentially when

[36] For an approach to this question see for example Cunchillos, C. Introduction à la théorie des unités de niveau d'intégration *in* Tort P. Pour Darwin.

[37] See for example Barabasi, A.L. and Olt Vai E.N. Nature Rev. Genet. 2004; 5: 101–13.

creating a new link so that a scale-free network is put in place. It could also be generated by a duplication model followed by gene divergence. The scalar character of biological networks leads us to consider a network as a fractal object, reproducing the same typology and thus potentially the same properties at different scales.

Biological networks also have the characteristic of being modular. This means that some nodes are interconnected more specifically, forming tightly connected small groups. These modules can then be considered (at least as a first approximation) as independent units. For example, proteins participating in the same metabolic chain or proteins contributing to specific signalling pathways may be considered as forming modules for a cell. The different parts of the nephron can be considered as the functional modules of the kidney. The large physiological systems (respiratory, circulatory, etc.) of the human body can also be assimilated to modules. The castes of insects too ... The advantage of this modularity is that it is then possible to isolate regions of the network behaving relatively autonomously. There are also development modules with possibilities for reversion, addition, subtraction, etc. These changes are explanatory of phenotypic variations by simple rearrangement of modules.[38] It is thus believed that neotenia (prolonged preservation of juvenile characters) is an essential distinctive element between man and chimpanzee.[39]

With the modularity of networks, creation is therefore essentially combinatorial in biology as in many other fields.

If a module is a unit of a network, it is also connected to the other modules by "intermodular" nodes. The network forms an indivisible whole, each module or character being separated only in a relative way. There is therefore a self-adaptation of the modules to each other, corresponding to the collective properties of the network and making it possible to maintain a global structure. The organization of the modules themselves can make modules appear on a higher organizational scale. It is even an expected property of scale-frees networks.

Thus, in the hierarchical scale of the living, at the lowest level, the nodes of the network are macromolecules interacting together. These molecules form structural and functional units or modules that are nothing else than the nodes of the cellular network. The cells also organize themselves into modules: tissues, organs, organisms that themselves form ecosystems and societies. At each hierarchical level, the modules at level i must be seen as functional units for level $i + 1$. It is their state of activity that defines the phenotype at the $i + 1$ level.

---

38 See West-Eberhard, M.J. Developmental plasticity and evolution.

39 It is enough to compare a photograph of an adult human and a young chimpanzee to be convinced. See also Chaline, J. and Marchand, D. Les merveilles de l'évolution, Chapter 15 or Gould, S.J. Ever since Darwin: reflection in natural history, Chapter 7.

Since the macromolecule and at each level of the hierarchical scale of the living, an elementary phenotypic character can then be assimilated to a coherent structural or functional unit participating in the realization of the living, that is to say a module of the network. And the value of the phenotypic character is then given by the functional state of the module. According to this definition, certain genes encoding macromolecules essential to the module will play a preponderant role. However, since the network forms an integrated whole with multi-level feedback loops, no gene can be considered completely neutral for a given phenotype.

In a network, causality is actually an irrelevant notion. Indeed, the presence of many feedback loops leads to circular chains of causality. Thus, if node 1 interacts with node 2 which interacts with 3 which itself interacts with 1, then it is impossible to say whether it is 1, 2 or 3 that modifies the activity of the network as a whole. Here we find again the notion of modules, that is, highly interconnected nodes acting as an integrated and coherent set. These circular causal chains are probably an essential reason why the causality between the genotype and the phenotype is degraded. Cases of less degraded causality should be represented by phenotypic traits (or modules) that are defined by only one node (or a major node). Thus, in the case of Mendelian diseases, the products of the genes involved seem to be often characterized by poor connectivity.[40] But points of less robustness of the network can be observed at all the hierarchical scales of the living: a specific enzyme or receptor unique for a cell (e.g., glycogenolysis enzymes), a unique cell type to perform a function (e.g., the hepatic cell and urea cycle), a unique organism for behavior (e.g., queens for the reproduction of bee societies) or a single species in an ecosystem (e.g., large carnivorous predators or pollinating insects).

The elementary phenotypic characters can be assimilated to modules of a functional network for the scale of the living being considered. The state of all the modules then defines the overall phenotype of the organism at a time t. The state of a limited number of modules can also define a complex character as a disease. We have been working with Jean-Marc Victor, Gaelle Debret, Annick Lesne, Leigh Pascoe and Gilles Wainrib on this issue regarding Crohn's disease considering the hypothesis of functional biological modules.[41] We can then show that a limited number ($n \approx 12$) of functional modules can explain the incidence curves of the human disease. Thus, the model proposed here can be applied in practice.

[40] Feldman, I. et al. Proc. Natl. Acad. Sci. USA 2008; 105: 4323–8.

[41] Victor, J.M. et al. PLOS One 2016.

4

The Role of Genes

Quantum mechanics is certainly impressive. But an inner voice tells me that it is not yet the last word. The theory has much to offer, but it hardly brings us closer to the secrets of the Old. In any case, I am convinced that He does not play dice.

Albert Einstein

The word "gene" was introduced by Johannsen in 1909 to define a "factor" transmitting a phenotypic character without making any hypothesis about its physical nature.[42] It was meant to be an operative word to reflect on the inheritance of the phenotype. With the discovery and decoding of the DNA molecule, the definition of the gene has changed and the term has, in its modern sense, lost its "pheno-centric" character. In its modern definition (which we use throughout this text unless otherwise stated), the gene is seen as a portion of the genome most often encoding a protein but also sometimes for a small RNA. It has been known for several years that a gene can encode several proteins by alternative splicing and that a protein can have several functions. It is known that it comprises regulatory regions, sometimes very remote from the coding part itself. There are also several promoters, overlapping or nested genes, etc. All this knowledge makes it difficult today to define the gene (in its modern sense) as a physical or functional entity.

The term "gene" nevertheless retains widely recognized operating qualities. Taken together, genes and non-coding parts of DNA form a sum of multiple coding, regulatory or unknown function elements. This set is

[42] For more details see Fox Keller, E. The Century of the Gene.

called the "genome". It corresponds to all the genetic information carried by DNA.

Genetic information is used for the ontogenesis of the living organism, giving it its shape, structure and functioning. After ontogeny, genetic information helps to maintain the homeostasis of the living at any moment of its life, by regulating qualitatively and quantitatively the possibility of synthesis of proteins. Genetic information is thus inseparable at all times from the phenotype. This notion is important to clarify at the outset because genetic information is often seen as a development program and it is implied that the phenotype is then stable and only depends on the genome marginally. In fact, the genotype/phenotype duality is verifiable at all times.

Thus, genetic information may be altered in some cases throughout life. In prokaryotes, recombination causes a bacterium to integrate an exogenous DNA molecule. Its behavior will then be modified and the bacterium will acquire new properties (especially resistance to antibiotics). This is Oswald Avery's classic pneumococcal transformation experiment. This property has been widely used by molecular biologists as the recombinant DNA technique. Genetic information can therefore be modified and lead to a significant change in the phenotype. In eukaryotes, the genome is usually stable throughout life. There is, however, one notable exception, cancer, where the cancer cell undergoes one or more somatic mutations. This modified DNA cell then changes its behavior, multiplies without control, invades adjacent tissues and then metastases through the body. Lymphocytes, cells of adaptive immunity, are another exception in vertebrates. Each cell rearranges its genes encoding the immunoglobulins or the T receptor to form a single protein. The millions of lymphocytes in the body then provide an immense repertoire of proteins capable of virtually recognizing any foreign biological structure. These examples confirm that genetic information can be updated and modulates the phenotype at any time.

At present, the genome's function is to provide information on the quality and quantity of proteins that the body can produce. It is therefore essential for the constancy of the synthesis of proteins and beyond that for their biological function. As these proteins determine the structure and function of cells, the genome is indirectly essential to cell functioning. As the cells of a multicellular organism are important to the life of the organism, the genome ensures the perenniality of the structure and function of the organism. In the end, the genotype is linked to the phenotype by a chain of causality that occurs throughout the hierarchical scale of the living. But at each bar of the scale, the genotype/phenotype relationship becomes looser and it becomes increasingly difficult, as one climbs the scale, to admit that this causal chain leads alone to a predetermined embodiment. This observation is even more true if we take into account the fact that the organism evolves in a changing environment and that there are important stochastic variations in the expression of genes or protein interactions, for example. The genetic information provided for the realization of the phenotype cannot therefore be strict but rather fuzzy. It must be seen as an indication of a set of potentialities (Figure 1).

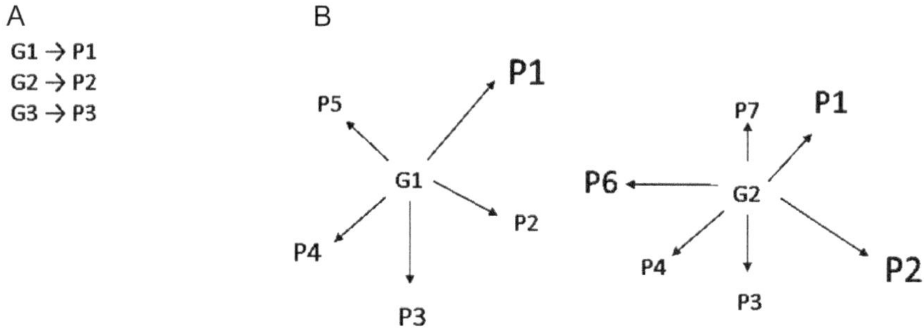

Figure 1. Mendelian and probabilistic models for the genotype/phenotype relationship. In the Mendelian model (A), the genotype/phenotype relationship is univocal: a given genotype gives only one phenotype. In the probabilistic model (B), several phenotypes (P1-5) can result from a given genotype (G1), each with a different probability (symbolized here by the size of the arrows and letters). For another genotype (G2), the field of possible phenotypes and their respective probabilities are different. G: genotype. P: phenotype.

It is the current triumph of molecular biology to inventory all the correlations between genetic variations and functional states of cells or multicellular organisms. This method of differential gene analysis abounds in scientific journals for several decades. However, all experiments in biology always include a statistical aspect because prediction is generally not possible for a given biological component. As expected for a fuzzy genetic program, there is always an intrinsic variability in life. Biologists have attempted to minimize this variability (most often seen as background noise) to reduce phenotypic diversity and access the chain of determinations from genotype to phenotype. They have created cell clones that are the stable lines of laboratories or lines of consanguineous animals that are extremely homogeneous from a genetic point of view. Despite this, understanding individual determinism is a goal that is usually considered inaccessible. It is therefore the study of a multitude of individual cases that evaluates the field of possible phenotypes and calculates a probability of observation of each one. The set of possibilities is accessible only by a statistical study.[43]

Classic genetics associates a gene with an enzyme and an allele with a phenotypic trait. A gene (defined in its modern sense as a nucleotide sequence necessary and sufficient to encode a protein) is then seen as a

[43] Statistics thus owe much to biology and genetics with the work of Fisher, Galton and Pearson for example.

quantum of hereditary information. For a given gene, the properties of the encoded protein can then be seen as the possible phenotypes of the gene. But if one ascends into the hierarchy of living things, there is no known gene defining the properties of biological units such as the cell, the organism or animal society. One can no longer individualize a quantum of genetic information for a trait. Genetic information becomes an indivisible whole and the elementary character is the result of the genome as a whole. It is then absurd somewhere to want to look for the genes of a given character, beyond the scale of the protein. It is the notion of the genome that matters at higher scales. It is increasingly common to consider the genomic dimension of information rather than genes taken in isolation. Thus, a gene (in its modern definition) can best define a phenotype only at the molecular level and therefore does not correspond to Johannsen's definition for higher scales of life. With the genomic dimension of information, the question of the gene in its "pheno-centric" definition of 1909 (i.e., a hereditary factor associated with a phenotypic trait) remains a topical issue beyond the protein.

Let's take an example. I love blue hydrangeas and I choose them for their color. Some hydrangeas have mutations of the genes encoding their pigments. But I can also add slate to the feet of pink hydrangea which will allow them to take the desired color. In this case, I choose hydrangeas with a particular metabolism. Finally, some plants can be more blue without adding slate but simply because they will draw the minerals more easily because they have deep roots. The plants with a particular rooting behavior will then be selected. Each time, we talk about genes that touch different functional modules: pigmentation, metabolism, root biology, etc. If one pushes the reasoning, all genetic variations of the plant can contribute more or less to the phenotype "blue color". A character is not defined by a single gene but by the whole genome. Thus, many genes (in the modern sense) contribute to the transmission of a phenotypic character. This is what is actually observed for diseases or complex traits as we saw in the introduction. But then, what is a gene in the sense of Johannsen? What information tells me that a hydrangea will be blue or not when I plant it? It seems that this is the whole genome for a given environment.

One would think that the idea that a phenotypic character belongs to the whole genome is incompatible with the observation of Mendelian characters. For a Mendelian trait, an allele alone determines the value of a phenotypic character. In fact, a Mendelian trait most often translates an impossibility to realize a value of the phenotype usually possible with a comparable genome. This is due to the defect or loss of function of the protein encoded by the mutated gene. If we take the example of hydrangeas, "blue" may be due to the absence of possible synthesis of the "pink" pigment. We are then in a particular case where for the color phenotype, the field of possibilities is amputated by one or more values by the mutation dealing with the single protein, the final path of the phenotypic character "color". The Mendelian mutation affects the most proximal cause of the phenotypic character that can no longer be realized. But character is not defined solely by this mutation since other causes can lead to the "blue" character.

If the genotype/phenotype relationship is loose, one can wonder about the size of the space of possible phenotypes carried by a genome.[44] Artificial selection experiments seem to show that the phenotype space is immense for identical or similar genotypes. The breeds of dogs are extremely different, from the Chihuahua to the Bernese mountain dog through to the greyhound. It is the same with cats or vegetables such as potatoes or tomatoes. Improvements in domesticated species seem endless as shown by selection of floral varieties or selection on a specific trait such as lipid composition of maize, wheat grain size, or milk production of cows. Thus, the number of possible phenotypic values for a genome (or set of nearby genomes) seems very large.

The genome of a chimpanzee and a human only differ by 1.5%.[45] This difference is probably not greater than that observed for the genomes of two varieties of rice. We can then ask whether man and chimpanzee are not possible phenotypes of (quasi)-identical genomes. This question may seem provocative, but it is only the logical consequence of the probabilistic nature of genetic information. To what extent can genomes be considered as different or identical and to what extent can different phenotypes be predicted? From what percentage of genotypic differences can we expect regular phenotypic effects? There is, in fact, no simple answer to this question.

We are in fact facing a paradox. On the one hand, classical genetics based on Mendel's laws indicate that a single nucleotide change over 3 billion can induce disease or death. Similarly, quantitatively modest genetic differences may define different species. On the other hand, Kimura's work shows that the vast majority of genetic variations probably do not have such a drastic phenotypic effect since most are neutral with respect to natural selection.[46] In fact, genetic variations accumulate in the genomes without any significant phenotypic modification of the species. The results of genome screenings for complex human diseases indicate that there are few effective polymorphisms. Genome sequencing studies confirm that most of the millions of identified genetic polymorphisms are likely to carry little or no phenotypic effect immediately detectable. A recent work on the sequencing of the genome of humans shows that a healthy individual carries on average at least one hundred mutations "loss of function" in his genes

[44] See the book by Wagner, A. The origins of evolutionary innovations.

[45] The difference between us and our closest cousins seems to be mostly due to variations in gene promoter regions.

[46] Kimura, M. Théorie neutraliste de l'évolution.

of which 20% in the homozygous[47] state. It must therefore be admitted that many mutations, including those which appear to have a deleterious effect on the encoded protein, often have little effect on the whole organism. We therefore have contradictory data that genetic variations of the same quality (mutations loss of function) can have a very strong effect or on the contrary be invisible on the phenotype. In addition, a morbid allele in one species may be the normal allele in another species. In total, from one individual to another, the same mutation does not have the same phenotypic impact. We find here the idea that the gene in the modern sense of the term (a DNA sequence), even Mendelian, does not completely predict the phenotype.

The phenotypic effect of a gene is readily studied by an approach of invalidation in a model organism (knock-out). In the most severe cases, invalidation is lethal to the organism or its offspring and the gene is classified as essential. In the C. elegans worm, fewer than 30% of the genes are lethal.[48] In some cases, a phenotypic effect can be observed which can be considered as a disease (i.e., a non-lethal anomaly). Often, the phenotypic effect is unapparent. In Arabidopsis, only a small fraction of the invalidated lines has a plant-wide phenotype.[49] In mice, 10% to 15% of the genes were invalidated. 10% to 15% of these genetically transformed mice have no apparent phenotype.[50] More recently, a systematic invalidation of murine genes has been developed.[51] Of 500 strains, 58% are completely viable in the homozygous state and 13% more have only a decreased number of homozygous offspring. Of the 250 lines of invalidated mice with highly detailed phenotyping, 35% had no identifiable abnormalities. In these models, the lack of phenotypic expression is usually considered as a proof of functional redundancy (or to be semantically rigorous of degeneration). It can therefore be deduced that many genetic variations, even drastic ones, do not significantly displace the range of possible phenotypes.

Modelling studies confirm that there are numerous genotypes capable of producing the same phenotype.[52] In these studies, the space of genotypes (respectively phenotypes) corresponds to the set of all the genomes (respectively phenotypes) that can be realized. Genotypes producing the same phenotype can be linked together in genotype space, that is, one can systematically find a chain of intermediate genotypes, separated by a single mutation between each of them, and all producing the same phenotype.

---

[47] MacArthur, D.G. et al. Science 2012; 335: 823–828.

[48] Kemphues. Worm Book. 2005 Dec 24: 1–7.

[49] Bouche et al. Current Opinion in Plant Biology 2001; 4: 111–117.

[50] Barbaric I et al. Brief Funct Genomic Proteomic. 2007; 6: 91–103.

[51] White, J.K. et al. Cell 2013; 154: 452–464.

[52] This paragraph refers to the work presented by Wagner, A. The origins of evolutionary innovations, Chapters 2 to 8.

This chain can sometimes be very long and the network of genotypes giving the same phenotype crosses a large part of the space of all the possible genotypes. A large number of genetic variations can therefore accumulate in a genome without phenotypic expression. In return, and as expected, the most probable phenotypes correspond to a greater number of genotypes among all the possible genotypes: these phenotypes are therefore more robust with respect to mutations. However, a single mutation can switch to a new phenotype for many genotypes. The genotype/phenotype relationship is therefore both robust (most often) but also occasionally very sensitive to point mutations.

In fact, these two properties are necessary for an innovation. The accumulation of modifications considered minor, by maintaining a stable phenotype, allows the preparation of the innovation without the lineage of the living being extinguished. This occurs only when an umpteenth mutation occurs that causes the old phenotype to switch to a new phenotype. The change seems brutal while it has been brewing in fact for a long time and this umpteenth mutation is only the straw that broke the camel's back. It is not necessarily more important than the others, it just happens last.

Thus, it must be understood that a Mendelian disease is only apparently due to a single mutation. It is due in fact to a whole series of genetic variations accumulated over generations by the population. The latter have settled over time and the population has become monomorphic for these mutations. But they are the ones that make the bed of the disease because they make the population as a whole subject to express the morbid phenotype in case of an ultimate mutation propitious. The disease seems to be determined by a single apparent cause (the Mendelian genetic polymorphism defining the disease) and our reasoning neglects the other genetic causes which are fixed, old and therefore unapparent for the geneticist who works on polymorphisms. Yet it is these accumulated mutations which define the propensity of the population to see the appearance of the sick. This point of view explains why the genetic variant morbid for one species is an ordinary genetic variant for another. This also explains why a homologous mutation to a human morbid mutation usually leads to no disease in the animal. Finally, we see that the Mendelian traits (morbid or not) are only a particular case of phenotypes whose determinism is always complex.

5

The Share of the Environment

We now know that the moon is demonstrably not there when nobody is watching.

N. David Mermin

The interaction of a genome with an environment is necessary to transform the organism from a state of becoming potential (the genome) to a concrete and unique state of realization (the phenotype). The phenotype of a living organism self-realizes when it interacts with a particular environment. It is in fact a simple and intuitive definition of creation: the construction of a new individual from its genotype when it encounters an environment. This is the case of a coconut that drifts on the sea and encounters an island, a dandelion seed carried by the wind that touches the earth, a seed of mistletoe deposited by an animal on a branch of an apple tree, a Tenia egg ingested with meat that settles in the intestine, a turtle egg deposited on a beach, a fertilized egg implanted in a uterus, and so on. The phenotype then unfolds by interaction between the genome and the environment.

The simplest phenotypic character is that of the conformation of proteins. A protein has a three-dimensional structure that is not unique. It depends on its primary structure precisely defined by the coding sequence of DNA (the genotype). But it also depends on the conditions of the environment: pH, salinity, presence of chaperone proteins, associations in multi-protein complexes, presence of various ligands, etc. It also depends on changes made by the cell (post-translational modifications, expression rate). All these "environmental" parameters for the protein condition its function and thus its biological role. We see that the phenotype, even on the most fundamental scale of the living hierarchy, depends on the genotype but also on multiple parameters related to its local environment. For a protein, the local environment is essentially defined by the phenotype of the cell to which it belongs. The cell forms the ecological niche of the protein. It is its relative ecosystem, the protein being separated from the rest of the world by the plasma membrane.

On the cellular scale, the phenotype also depends on the genetic information on the translated proteins. This genetic information is in particular that located in the promoters or the regulatory regions of the genes. But cellular phenotype is highly dependent on non-genetic parameters: epigenetic mechanisms that regulate the expression of genes, environmental factors (hormones, exogenous substances, nutrients, various stresses (heat, toxic, radiation, etc.), cellular contact, etc.). In fact, from the same genome, there is considerable cellular plasticity, especially in eukaryotes. Thus, there are 256 cell types in humans that are as different as neurons, enterocytes, white blood cells, platelets, spermatozoa, ova, and so on. The cell phenotype is complex and therefore certainly less directly dependent on genetic information itself than the molecular phenotype.

The cell, when isolated, is an organism in its own right (case of bacteria or monocellular eukaryotes). When it is integrated into an animal or plant pluricellular organism, its environment becomes that organism which constitutes its ecosystem and protects it more or less from the external environment, just as the cell constitutes the ecosystem of the protein. This protection or isolation of the external environment can be marked when the organism has developed a stable internal environment as in mammals or birds. It can be less stable as in plants, sponges, jellyfish, etc. In the case of a stable internal environment, the environment becomes more fixed for the lower hierarchical levels of the living (cell, molecule).

It can thus be seen that at all scales of life the phenotype P is the resultant at all scales and at all times of the environment E and of the genetic information G. We usually speak of the gene-environment interaction (or reaction norm) GxE which defines the phenotype of a given individual.[53] We can then say that the phenotype P is a function of G and E: P = f (G, E). The part of genetics and the environment may be different from one phenotypic character to another. Thus, hair color is a phenotype strongly predicted by genetics while their length is a phenotype very dependent on the environment in man (but also in woman!).

Experiments carried out in recent years have shown that the phenotype also has a stochastic or random component. Thus, the rate of expression of the genes from one cell to another, under the same culture conditions, is variable. Protein-protein interactions are not as specific as imagined with the classic allosteric model. The interactions of proteins with chromatin also vary considerably. It is therefore necessary to add to the two components

[53] For further discussion on the reaction norm see Lewontin, R.C. The triple helix, Chapter 1.

G and E which define the phenotype, a random or stochastic component. It can then be written that the phenotype P is a random variable of the function f (E, G) (Figure 2). This amounts to considering GxE reaction norm as a point cloud rather than a linear function.

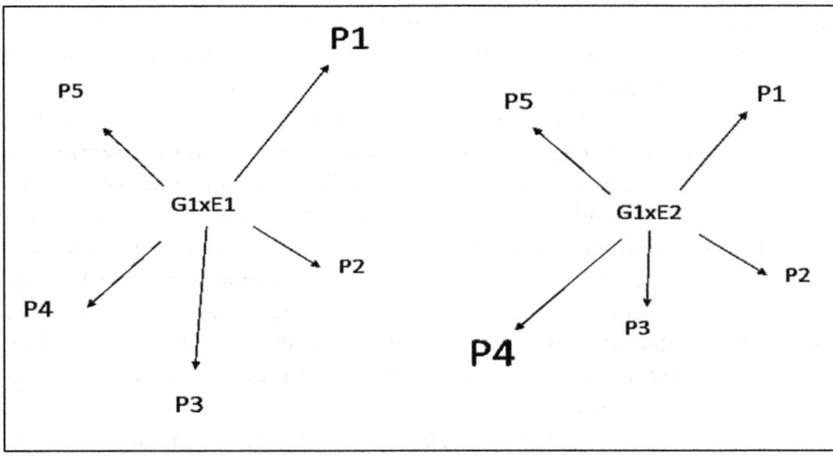

Figure 2. For the same genotype G1, the environment (E) modulates the probability of the possible phenotypes (P).

As indicated above, the environment for the phenotype of proteins or cells is often the phenotype of the cell or organism to which they are attached hierarchically. There is therefore a system of hierarchical dependence of the biological units that are nested as Russian dolls. It is then possible to define a correspondence between two nested systems (the multicellular organism and the cell in a multicellular organism, for example). If we consider the ecosystem in which the organism lives as its complete environment, we can even generalize this formula and write for each level of integration of living i, that P_i is a random variable of the function f (P_{i+1}, G) with $i = 1$ for the molecular level, the maximum value of i corresponding to the state of the ecosystem in which the whole organism lives (cell for bacteria, individual for multicellulars, animal society for eusocial animals, ecosystem for species, etc.).

Some traits appear very early in the development (e.g., the placement of the three embryonic layers of the ropes), others are late as the transformation of the butterfly, the formation of seeds in the plant or the age of puberty in the human being. These late characters are, by definition, indeterminate until their realization. If the characters appear only sequentially, then the later characters may be partly conditioned by the earlier characters. The formation of the central nervous system is, for example, dependent on the placement of the embryonic layers. The ontogenesis can then be seen as a cascade of characters, which corresponds well to our knowledge of embryology, which consists of a cascade of information exchanged between inductive cells and competent cells. There is therefore a hierarchical structure over time also.

Each character is thus defined by the interaction of the organism with its environment. The environment modulates the probabilities of realization of the genotype (Figure 2). It actively contributes to defining possible phenotypes. And if self-realization of the phenotype takes place according to its environment, then the ecosystem contributes to inducing the phenotype of the organism which contributes to the realization of the cellular phenotype and in turn induces the molecular phenotype. We therefore have a cascade of consequences that regulates the phenotype throughout the hierarchical scale of life. This cascade of determinations is inverse to the cascade carried by the genetic information which starts from the bottom of the scale to raise the bars. It is consistent with the observation that the molecular scale is more genetically determined than the higher scales of organization and brings some symmetry to the relationship between genotype and environment.

Applied to ontogeny, the self-realization of the organism is a sequential process with cascades of induction. The genotype depends on the genome and the environment (essentially biochemical gradients) from the early stages of division of the egg. On this basis, the layers, the organs and then the tissues, etc., will be placed. This cascade of events is linked to induction. In contrast, as long as a character is not realized, it is indeterminate (and only approachable by the statistic). This is the case of the age of puberty in childhood or that of menopause in adulthood, for example.

Finally, it should be noted that the phenotype is not fixed. It is constantly mobile. The organism self-realizes itself from its genome and in its environment at every moment. This point is often overlooked because the notion of character has been widely used for the classification of species. It is easier to classify phenotypic elements considered fixed (bone form for example) than from variable elements (serum cholesterol for example), this is particularly true in paleontology where, by nature, the characteristics studied are permanent and fixed. But the phenotype is naturally changing, reflecting the adaptation of the living organism to its environment and to random phenomena. At every moment, a human being modifies his heart rhythm, his breathing, his vasodilatation, and so on. Even for characters considered sufficiently fixed to define a species, there are significant variations. We can cite the color of the plumage of seagulls in our regions, which changes according to the season; the shape of the bones that depends on diseases or aging; the wearing down of teeth that depend on food, etc.

Beyond ontogeny, the phenotype remains adaptable and potentially changing. The living organism is "realized" at all times. Without repeated adjustments, there would be no possible adaptation to the environment of living beings. In general it is judged that an organism adapts, but one could also say that the environment modulates it. Thus, to say that a cell adapts to the conditions of the organism or that the body controls the state of a cell is only a question of a point of view.

It is impossible to set a biological parameter without leaving other parameters free. Thus, for example, adapting the body to running means leaving the heart rate free to move. If, on the other hand, the heart rate is set at a limit (for lack of training for example), then it is the pace of the race that will be forced to adapt. It can be seen that the determination of each phenotypic character can only be temporary and sequential. A phenotypic character thus takes on a certain freedom when another character is temporarily fixed. This way of thinking corresponds to a network of interactions with permanently regulated flows and adaptations of the solicited nodes step by step. It merely translates the view that living beings are based on a biological network.

Paradoxically, it must be understood that the "realization" of the phenotype in its environment cannot be continuous but must be sequential. The environment is changing and the organization must adapt to it at every moment. To be able to change, the phenotype can then take only momentary values. Between two phenotypic determinations, the organism must then have an area of freedom, in other words, a certain indetermination, if it is admitted that it can respond to the environment of the next instant. The phenotypic determination by the environment is therefore necessarily sequential for the phenotypic adaptation to take place.

The sequential nature of the phenotypic adaptation implies that, for even minimal time intervals, the organism regains the potentiality of all phenotypes compatible with the genotype. One could therefore imagine that it is completely different at time t1 with respect to the instant t0 which precedes it. We would all be transformers, those robots that change from one moment to another. This theoretical freedom of the phenotype is probably restricted by the fact that the environment exerts its action at very short intervals, thus forming determinism, but it should be seen in spite of everything, at least in an exceptional way. Yet, in reality, the phenotype can be considered as fixed over long periods for many parameters of the living, giving the impression of a continuity of states. This is one of the criteria listed for the definition of living organisms. It is then conceivable that other factors are necessary to stabilize the phenotype between two measurements.

6

Analogy Between Organisms and Particles

I tried to explain to these philosophers the interpretation of quantum theory. After my presentation, there were no objections or difficult questions; But I must admit that it was precisely that which shocked me most. For if, at first glance, one is not horrified by quantum theory, one has certainly not understood it. Probably my presentation was so bad that nobody understood what it was about.

Niels Bohr

The genotype/phenotype duality is intrinsic to the issue of life. There is no modern living organism without a genotype/phenotype binomial. A genotype does not have its own reality, it must be carried by a cell already alive to reveal itself. It is only virtuality. Conversely, no living being can emerge ex nihilo without a genome to provide the information necessary for its ontogenesis and the maintenance of its integrity throughout its life. Without genotype, today there is only an unorganized mass. A living organism is therefore defined by a duality that unites its genotype and its phenotype. And if there is no direct relationship which can be extrapolated between genotype and phenotype (i.e., program), the organism cannot be reduced to only one or the other of its two components. It is therefore necessarily defined both by one and the other. This situation recalls the known wave/particle duality in quantum physics.

A century ago, physicists, most of them very young-except Planck (Einstein, Bohr, Born, de Broglie, Schrödinger, Dirac, Heisenberg ...)

invented quantum physics.[54] Quantum mechanics is concerned with the study of microscopic systems (atoms, particles). On this scale, a system has paradoxical properties and to describe it, it is necessary to represent it in two complementary aspects: an undulatory aspect and a corpuscular aspect. The classic experience of the two slits summarizes this duality.[55] To solve this paradox, these physicists proposed a new formalism. This formalism is, by definition, only a representation of reality (whose intrinsic nature is by definition impossible to apprehend by science alone[56]). It is, moreover, difficult to reconcile with our intuition of the world, and the duality of the wave-corpuscle has made (and still is) the subject of many philosophical interpretations.[57] Nevertheless, quantum formalism has never been faulted in explaining the world on a microscopic scale. It has proven to be fully operational in its calculations and predictions and has enabled major scientific and technological developments that are part of our everyday life (lasers, electronic components, etc.).

The state of a microscopic (or quantum) system can be specified by several mathematical tools: a wave function, a matrix, or a state vector. The wave function (or matrix or state vector) summarizes the properties of the system (resulting from the physical preparation that generated it) and defines its probabilities of being in one state or another at a given moment. The wave function carries information on both the nature of the particle and its physical environment. The corpuscular nature of the particle is revealed only during the measurement of the particle, that is, when the particle is interacting with an instrument. Therefore, before the physical measurement of the system, the particle is considered as in a superimposed state of all the possibilities defined by the wave function.[58] The measure confers on the particle a status which is no longer indeterminate. In a way, it leads to

---

[54] It is irrelevant here to review what quantum physics is. The interested reader will find the references of several works in the bibliographic section.

[55] The experiment of the two slits is reported in all the works of vulgarization of quantum physics. We refer the reader to these works for a detailed description. Briefly, in the experiment a beam of particles (photons, electrons or other particles) is projected onto a plate pierced with two slits and their arrival position is detected on a screen located behind the plate. Instead of detecting the particles in direct view of the two slits, lines are observed as if the beam were a wave diffracted by the slits and the particles followed a path predicted by wave physics. Yet particles are corpuscles that interfere with the screen in the form of dots. Moreover, the experiment gives the same results if the particles are emitted not in bundles but one by one. For some physicists, this experience of the two slits sums up the very essence of quantum physics.

[56] "We must remember that what we observe is not Nature itself, but Nature subjected to our method of questioning". Werner Heisenberg.

[57] For a detailed discussion see Laloe, F. Comprenons nous réellement la physique quantique?

[58] The superimposed character of a microscopic system before the measurement is explained by the Schrodinger's cat's paradox.

the choice of a single solution among all the possible states (the reduction of the wave packet).

The gene-environment interaction GxE can be seen as analog to a quantum wave function (Figure 2).

One could object that the genotype possesses a materiality in the form of, one or more, long molecules of DNA, contrary to the wave function or to the state vector of quantum physics which have no materiality. It is, however, by its informational nature and not by its physical nature that the genotype reveals its importance. It is also the informational character of the quantum wave that is important in physics and some authors even raise the question of the essentially informational nature of the quantum wave.[59] In fact, this one carries information on the possible states (and their respective probability densities) of a microscopic system (energy, speed, probability of its position in space...). The probabilities of realization are equal to the square of the amplitude of the wave function for a place and a given instant.

The measurement of a particle always involves experimental equipment and a detection apparatus. The experimental system also participates in the definition of the wave function. Thus, for example, in the Yung slot experiment, the number of slots changes the final result. We have seen that in biology, the environment also participates in the definition of probabilities of realization. The sensing device also influences the measurement. It is indeed impossible to take a measurement without interacting with the system studied. For example, to know the position of an electron, it must be detected with a photon which immediately transmits energy to it. Thus, the state of a particle is known only by the modification which is brought to it by measurement. It is always necessary to consider the macroscopic global system as "particle-experimental system". There is no absolute "neutrality" to the extent and reality is only known after it has been manipulated and thus already modified. By analogy with the quantum wave function, the genome can only be revealed in an environment that contributes to the realization of the phenotype. Thus, it is not the genome alone but the genome in its environment that carries the set of all possible phenotypes and which must be considered analogous to the wave function.

As with quantum physics, the relationship between the genotype and the phenotype is probabilistic in nature. The genotype and its environment together define the "space of possible phenotypes" as the wave function defines the set of possible quantum states for a particle in a physical system. The phenotype is the realization, unique among all the possible ones carried by the information gene/environment, during the creation of the living being. It is an achievement, in the true sense of the word, that is, the projection of a virtual object into the real world. It is analogous to the

[59] Laloe, F. Comprenons-nous vraiment la mécanique quantique? p. 220.

reduction of the wave packet during the measurement. The link between genotype and phenotype can therefore be seen as analogous to the link between the wave function and the measured state of a particle.

In quantum physics, the measure of a preparation is made by the experimenter. However, current theories on decoherence suggest that in Nature (characterized by the absence of a quantum physicist within the range of experience!), if a meter is not met, a microscopic system interacts with its environment, which is enough to reduce the wave packet and to cause a measure, "in real life". The natural environment of the particle serves as a measuring instrument. The analogy with biology is then easy to make. The physical measurement and reduction of the wave packet are analogous to the creation of the living being. As in quantum physics where the measurement ends the experiment, the development of the living organism fixes it in a phenotype. Given the large number of possible genotypes and the immensity of possible phenotypes, it is unlikely that the same organism will appear several times (except in the particular case of monozygotic twins). The lookalikes are therefore exceptional (there is also a non-clonality theorem in quantum mechanics).[60]

The experimental situation of biologists is also comparable to that of quantum physicists. The physiological state of a cell or organism is known only during the experimental measurement, just as the state of the particle assumes a defined value only during the measurement. In both cases, the measurement value is unique and belongs to a set of possible values defined by the genotype and its environment (in the case of the cell) or by the wave function (in the case of the particle participating to a given system). Moreover, the gene-environment reaction norm GxE, like the wave function, carries information on the expected frequency of each value during the measurement. In both biology and physics, probabilities are accessible experimentally by multiple measurements on a large number of organisms or particles. Biology as quantum physics measures propensities.

In the formalism of quantum mechanics, a measurement operation (i.e., obtaining the value of a physical parameter such as position or velocity) is represented by an observable. The observable is a mathematical operator that gives the possible values that the parameter can take from the wave function. We can then consider that an observable is analogous to an elementary phenotypic trait. In quantum physics, it is the wave function as a whole that carries information on the observable and it is useless to seek to find "sub-functions" that would be attached to a specific observable. By analogy, we have seen that the whole genome participates in all phenotypic characters and no elementary phenotypic character is reducible to one or a few genes, just as the measurement of an observable depends on all the wave function.

---

[60] Laloe, F. Comprenons-nous vraiment la mécanique quantique?

There are, however, particular cases where an observable O can take only one definite value. We call this the proper state of the observable O. This eigenstate recalls what we observe for a Mendelian character in which the phenotypic trait (analogous to the quantum observable) can take only one perfectly predictable value. We then find again the idea that Mendelian genetics should be seen as a special case of the genotype/phenotype relationship.

If each elementary character (i.e., non-reducible to simpler characters) can be assimilated to a dimension of phenotypic space and treated as "separated observable" according to the expression of quantum physics, the overall phenotype of the organism is then a set of observables (size, weight, eye color, etc.). It can be defined by a space with N dimensions corresponding to the N characters considered as elementary. N appears to be large not only in complex organisms but also in bacteria, that is, in all hierarchical scales of life.

In quantum physics, an observable measured on a particle has a defined value only until the measurement of another observable (which does not commute) is made on the same microscopic system. The observable measured initially again becomes probabilistic at the very moment of this second measure. Observables that do not switch (called incompatible) correspond to non-independent physical parameters: the measurement of one implies the indetermination of the other. This phenomenon recalls the constraint of leaving certain phenotypic parameters free to allow to stabilize others in the case of external or internal variations to which an organism is subjected. On the contrary, there are complete sets of commutable observables (ECOCs) that can be considered as associated with independent parameter measurements. This concept agrees with that of biological modules, of which "states" can be considered relatively independent of one another. As an example, we have made this hypothesis for the modeling of Crohn's disease (see above).

In general, and finally, if the analogy between particle and organism is valid, then we should be able to apply the mathematical tools of quantum physics to the theory of networks. We have seen in the previous chapter that the phenotype could be understood as a biological network multi-scale. If the latter is also a set of observables, then the network itself can also be seen as a wave function whose states of the functional modules are possible observables.

7

Phenotypic Plasticity

Nothing is permanent except change.

Heraclitus of Ephesus

The adaptation of an organism to its environment is often assimilated to its phenotypic plasticity. The phenotypic plasticity refers to the plasticity of a material which is its property to modify its form under the effect of an action and to keep it at the stop of this action. According to this definition, plasticity also implies a fixity of the phenotype after the effect of the environmental factor has been exerted. We also speak of polyphenism suggesting that the living organism can take multiple different phenotypes depending on the environmental conditions. The plasticity of living organisms reflects (i) the fact that there are multiple possible phenotypes for the same genotype and (ii) the fact that the interaction with the environment is not neutral and that it induces a change in the self-organization of the living. Figure 2 is an illustration of such phenotypic plasticity.

However, a living organism is subject to rapid adaptations to various situations to respond to changes in the external environment. It must therefore permanently modify its functional network, that is to say its phenotype. This plasticity is more like that of a modeling clay that modifies its shape on demand, rather than a situation of choice among a more or less significant number of stable phenotypes. In this sense, it shows not a plasticity which suggests fixed forms more or less numerous but of a malleability which would therefore be a better term. We must imagine the living being capable of flexibility throughout his life. This is possible thanks to the biological network structure of living organisms. In the case of a network, it is permanently redetermined to adapt to external demands (external stimuli) or to changes in the network itself (breakdowns for example). Thus, the living material is permanently remodeled. Malleability testifies to a continuity in the global phenotype that adapts to each moment. Organisms can take all possible phenotypes at any time. They are like "barbapapa", these cartoon characters changing in each story.

According to the classical idea of phenotypic plasticity, we have seen that the acquired phenotype remains fixed in one configuration or another after the action of an environmental factor. With malleability, it is permanently "another", with no fixed continuity. Elasticity refers to the property of a material which deforms and then resumes its initial shape when the action ceases. It is therefore probably closer to what is observed for the living being that keeps a certain permanence of its phenotype while being permanently malleable.

Elasticity, however, implies the notion of the original form which serves as a reference point. For the living being, such an original form has meaning only if one accepts the notion of predetermination of the organization of the network. On the contrary, the genotype/phenotype relationship as developed so far implies that there is no phenotype without exposure of the living organism to its environment and that this phenotype is probabilistic. There is therefore in principle no reference organization that would refer to the concept of strict living programming. One can, however, take the phenotype already realized and put in place by ontogeny as the reference phenotype for the organism. In this case, the living person tends to return to this state of reference. This corresponds to what is called his homeostasis. It is sort of resiliency or a buffer effect. For a given organism, this reference state can be seen as a "resting" state of the biological network analogous to the state of least action in physics. In our development up to now, this state is therefore specific to each organism and is not predetermined. It corresponds to permanent information on the organization of the network.

At each hierarchical level of living organisms there is a biological unit which is self-organized with a network structure (structural domains of a macromolecule, a network of proteins and organelles for the cell, a network of cells and organs for an organism, caste network for an animal society, etc.). Network self-organization forms the basis of malleability. Through feedback loops, the network also has some resilience that brings it back to its previous state of lesser action after the end of the stimulus. The network is finally robust. The phenotypic robustness is the fact that if G and/or E move, P does not move. In other words, genotypic or environmental variations have no phenotypic effect. They are buffered. This robustness is a property of networks, especially scale-free networks that bear random breakdowns quite well. Robustness then makes it possible to accumulate genetic variations in the same population. On the other hand, the scale-free network is very sensitive to breakdowns on the most connected nodes. Finally, robustness can be seen as homeostasis when it comes to environmental variations. It is close to the concept of elasticity discussed above. We see here that knowing the properties of networks is essential both to understand the living and to analyse the concepts of plasticity, malleability, elasticity, etc. The concepts of robustness or elasticity make it possible to explore the permanence of the structure of the network, i.e., the permanence of the phenotype during life.

Over time, living organisms have evolved towards systems that are less and less dependent on their environment, setting characteristics such as, for example, body temperature, chemical composition of the internal milieu or protection of the fertilized egg. This evolution is the result of an evolving process of network self-organization. For this, it was necessary to define modules whose state is stable to the detriment of others, more labile which adapt around these more fixed modules. Thus, the secondary structure of a protein or early stages of development appears to be relatively stable in one species or even between species. This is partly due to the fact that the establishment of a hierarchical system has set environmental parameters for the lowest scales of the living hierarchy and for the early stages of ontogeny, thus limiting their phenotypic variability. The first stages of the construction of the living form then appear more rigid, allowing a more solid foundation for the construction of the upper floors.

We have seen that phenotypic plasticity as usually defined in genetic literature is the fact that at constant G, several values of P are possible if E changes. We also speak of polyphenism expressing the idea that a genome can be at the origin of several stable phenotypic values. Phenotypic plasticity, in the usual sense of the term, then indicates that there is a channeling of the phenotypes towards a small number of predefined possibilities. Organisms with closely related genomes form easily recognizable species, social insects form a small number of different castes, cell types are limited in number in a multicellular organism, the stages of development of insects or batrachians are also few. There are only two sexual phenotypes, and so on. The phenotypic plasticity defined in this way thus indicates a discontinuity of the phenotypes with fixation of a small number of phenotypes specified for a given genotype and environment. It is more a predetermination than adaptability. We will come back to this.

8

Behavior

A living system is an open yet stable system. It can be compared to a flame.

Léon Brillouin

The living organism can therefore be modeled as a self-organized network characterized by permanent state changes that reflect its adaptation to the environment, its tendency to homeostasis or simply stochastic changes in its functioning.

In physics, the movement of an object is characterized by its speed (in $m.s^{-1}$). Analogically, this speed can be compared to a stream in a network link. Knowing the mass of an object one can define its momentum (in $kg.m.s^{-1}$). Analogically, for a living organism modeled as a network, the amount of motion can be conceived as the set of fluxes passing through the biological network of the organism. The reference momentum can be considered as that of the "at rest" network, that is to say, its state of equilibrium or of less activity. This state of rest is characterized by a spontaneous optimization of the flows in the network.

To modify the amount of motion of a physical object, a force or action (expressed in Newton or $kg.m.s^{-2}$) must be applied to it. This force is usually seen as an external event that transmits an amount of motion to the object. For a network or a module of the network, this force can be assimilated to an input of which the result is a modification of the flows in the network. For a biological network, the result of this modification can be compared with a behavior, that is to say a reorganization of the flows passing through certain nodes or modules, defined themselves as the elementary phenotypic characters. If for a physical object, the force can come from a field or a direct interaction with another object, for an organism, the behavior can be dictated by a more or less durable environmental parameter (e.g., sunshine for synthesis chlorophyll by the plant, the presence of lactose in the culture medium of the bacterium for the adaptation of the metabolism of the bacterium) or by a punctual event (e.g., the encounter of prey and a predator).

Energy is an ability to transform a state.[61] Its size is in Joule (kg.m^2.s^{-2}). It is essentially a potentiality of action in the form of force. Thus, kinetic energy is the energy accumulated during the setting in motion. Potential energy is the energy available if an object is allowed to join another more stable equilibrium value. In general, the use of energy involves a transfer from one object to another or its use to perform an action. It is therefore a kind of exchange value or storage comparable in this sense to money. It can be stored and used later. It can be exchanged from one object to another. It can finally be used to "buy" mass, movement, a change of state, temperature, chemical bonds, etc. Due to the law of conservation of energy, everything that is realized or changes in fact corresponds to a transformation of energy and the object of physics is, to a large extent, to understand these transfers of energy (which brings physics and the market economy much closer!).

In living organisms, the change of state is often dissociated from the direct transfer of physic energy and matter. For example, if a fox sees a rabbit (or vice versa), the information received triggers a chase but does not provide the energy necessary for contraction of the muscles of the fox (or rabbit). The change of state of the animal is due to the interaction but only to its informational part. The energy used for the change of state is provided by the animal itself which uses its energy reserves (i.e., its potential energy). Living organisms can therefore respond to a signal without it being necessarily associated with a coupled transfer of matter and/or energy, which distinguishes them from most inanimate objects. The perception of the environment thus has the same consequences on the behavior of the living organism as a force or a transfer of energy on a physical object. A living object can then be defined as a system capable of processing, in its own way, information about its environment and modifying in response its functional network.

Living organisms can be receptive to many signals. In fact, living things detect objects, smells, flavors, textures, noise, ultrasounds, light waves in a wide spectrum of colors, polarization of light, terrestrial magnetism, temperature, duration, changes of position, personal states and those of other living beings, etc. Depending on the tools available to them, living organisms (including humans!) have a subjective apprehension of the world, defined by the word "umwelt" or "own world" by Von Uexküll. This own world in which the organism lives is its apprehension of the environment

---

61 https://en.wikipedia.org/wiki/Energy.

both in terms of perception (inputs) but also in terms of its reactions (outputs). It is very different from one living organism to another and thus leads to a great heterogeneity in the approach of the same environment by different organisms. Thus, the apprehension of the space of a fly is 50 cm, the sensitive world of an earthworm is made of vibrations and tastes, the proper world of a paramecium is associated with its eyelashes that enabling the recognition and avoidance of material objects.[62]

If the own world of a living organism is its specific way of recognizing and treating the information it perceives from the outside world, the living organism is also capable of perception of its "inner" performance, measuring most metabolic parameters, body position in space, organ performance, and so on. The internal information is thus treated, which is the basis of the adaptations between modules in the biological network. In reality, this internal perception and the way of responding to it corresponds to the "own worlds" of the lower hierarchical levels of life. Thus, the presence of a ligand participates in the world of the receptor, the presence of a hormone participates in the world of a given cell type, and so on. A complex organism is therefore made up of multiple biological units interlinked with each other and having very different worlds of their own. Here we find again the notion of a network where each biological unit corresponds to a functional module characterized by its inputs and outputs which themselves serve as inputs for other biological units.

In physics, the notion of energy implies a certain vagueness. Just as with the same amount of money one can buy a bike or open a savings account, energy is somewhat pluripotent. In the case of an interaction between objects, however, the transmitted energy is not always freely usable. Work expresses this notion well. It is the application of energy to an object by a force exerted over a certain distance (imagine yourself pushing a broken car). The work involves the use in a defined sense of energy. Thus, although it is expressed in Joules as pluripotent energy, it corresponds to an energy used for a precise "realization". To illustrate this notion, let's take the example of money that symbolizes energy. If I give my son a euro to buy a baguette or a euro of pocket money, the financial transaction is the same but my son knows exactly the difference between the money to do what he wants and the money to do something determined! Work (and by extension force or action) therefore implies a kind of constraint associated with the transfer of energy.

As with the use of energy, the processing of information by the biological unit may be more or less constrained. However, in general, the biological module that receives the input and returns an output is an "inhabited" box. Thus, instead of a strict determinism which characterizes the collision of

[62] The examples cited in this chapter are taken from the book by Von Uexküll, J. Milieu animal et milieu humain.

two macroscopic objects in classical physics, the living organism has the plasticity of response to its own environment. This plasticity seems all the more marked because the biological unity is capable of a complexity in its perception (i.e., integrating multiple stimuli) and in its actions (i.e., by integrating multiple sequences). Thus, the tick spends years waiting on a branch for the passage of a warm-blooded animal that it identifies with the presence of butyrate in the air. It then falls on the animal, recognizes its heat and stings for sucking blood. The tick is of interest only for some parameters of its environment (temperature, the presence of butyrate ...) and has few responses in return in case of stimulation (falling, crawling, pricking). His own world appears poor. On the contrary, a fish has a much wider range of perceptions and possible answers, signing its complexity.

This measure of the wealth of the own world can be extended to all hierarchical scales of life. The medium of an enzyme is limited for perception to its substrates and its products, to some regulatory elements, to the pH and to certain ionic concentrations. It is reduced to catalysis for its reactions. It is actually poorer than a bacterium that recognizes many other parameters and can respond in many ways to its environment.

The richness of the world proper to a biological unity is thus partly linked to the complexity of its network. This richness can lead to a wider range of behaviors for a given input. This behavioral variance can be assimilated to the free will of the biological unity. This is largely due to its complexity. However, free will can be limited even among complex organisms. Thus, the own world of a tick appears very limited despite the relative complexity of the animal. In fact, the body's response capacity is largely hereditary and stereotyped. We speak of instinct. Instinct then acts as a mechanism that reduces the behavioral variance of biological units. Its advantage in return, according to the term of Uexkull, is to "secure" the own world of biological unity. By channeling behavior, it reduces the risk of inadequate response of the biological unit and thus loss of its lineage (for an organism) or disease (for a biological unit integrated into a larger organism).

We see here that the environment consists not only in molecules or energy used to build the organism but also, in large part, as a source of information. The organism can therefore be seen as a machine to process information to arrive at a self-measured and self-realized structure. And since it is the perceived share of the environment that has an essential role, it can be said that the phenotype largely results from the reaction norm "GxE_{umwelt}" at all scales of life.

Behavior brings the organism from state A to state B. It relates to the biological unity as a whole, whatever the scale at which one places oneself to define unity (molecule, cells, organs, organisms, society). It is fundamental for the living. It ensures its adaptation at every moment to the variations of the external and internal environment. It is, however, something more than the basal phenotype since it is the way to change the flows in the network in response to information. It belongs to the dynamics of self-adjustment of the network and no longer to the value taken by it at a given instant.

These remarks on behavior naturally lead to an analogy with the quantum of action (and its minimal value, the Planck constant). An electron changes its state because of an external stimulus (absorption of a photon) or on the contrary interior (the emission of a photon). The change of state concerns the electron as a whole. For an electron around a nucleus it can result in an orbit change that is characterized by the emission or absorption of a photon of a defined wavelength belonging to the spectral band of the chemical element in which he participates. This action of the electron can therefore be considered as a behavior of the particle which leads to its change of state.

The principle of indeterminacy of Heisenberg indicates that it is not possible to know exactly, in a single measurement, both the momentum and the position of a particle. By analogy, we can say that it is impossible to study at the same time an organism at rest (a network fixed at a time t) and in action (an active network capable of reorganizing its flux). Indeed, knowing precisely the state of an organism (a cat for example) implies immobilizing it and even possibly killing it to dissect, weigh and analyse it chemically. Under these conditions, no change (adaptive modification of the organism) can be observed. On the contrary, the analysis of a behavior implies letting the cat do what it wants, without locking it in a cage. Ethologists, especially primatologists, know perfectly well that unhindered observation of animals in their natural environment is the best way to know their behavior. All in all, wanting to describe a cat in all its dimensions involves describing it both dead and alive!

The principle of indetermination refers to the knowledge that one can have of a particle during its measurement. But according to Bohr, it also testifies to the duality of wave-particle, since the measurement can only explore one or the other aspect, according to the experimental protocol implemented. Let us see what it is for living organisms.

We have seen that for a given organism, its biological network is organized to ensure the constancy of certain biological parameters. Thus, for example, the temperature or blood glucose levels are perfectly regulated in mammals. This fixed biological parameter implies a great flexibility of the regulating parameters (muscular contraction and sweating for thermoregulation, insulin and glucagon concentrations for blood glucose, etc.). It is seen that the fixation of a phenotypic trait implies a greater adaptability (and hence a greater variability or indeterminacy) for the phenotypic traits that are linked to it. Indeed, in a network, it is not possible to fix the flux passing through a node without forcing the neighboring nodes to dampen the flow variations. The living being must therefore choose around which essential parameters he must self-organize. Characters that are most stable imply the variability of others. We can then distinguish a hierarchy of traits (or modules) according to their fixity. Indeed, the characteristics of an organism are either constrained (and then not very adaptable) or adaptable (and then little constrained).

A living organism is then characterized not only by its molecular and cellular network and by its hierarchical organization but also by its points of strongest fixity and of greater phenotypic variability. These dynamic qualities are added to the static description of the network. They contribute to the phenotype and are shared by organisms belonging to the same lineage or taxon. These qualities are not directly genetic or environmental. They relate to a way of reacting to external or internal information, rules of global response of the network. They correspond to the characteristics of the "own world" of the organism. They represent determinism in the self-organization of the states of the network modules, some being considered more constant and others more variable.

It is difficult to understand what biology means to adapt an organism to its environment. However, considering the living organism as a self-organized network around more or less defined points of fixity, we understand that adaptation must be the bringing into play of the labile elements in order to maintain the points of fixity. The permanence of the living organism does not mean the stability of all the biological parameters but only the permanence of certain phenotypic traits characteristic of an individual. If the definition of the most labile or constrained modules is hereditary, this information may become characteristic of a species or a taxon. It then constitutes the heredity of the network self-organization.

Part 2
Heredity

Heredity is the transmission of the phenotype from one generation to the next. It is translated (and evaluated) by the excess resemblance between related individuals by comparison with the resemblance between unrelated individuals. It is usually summarized as the transmission of an individual's genetic information to their offspring. There are, however, other less studied forms of heredity, sometimes grouped under the term "soft heredity".[63] Thus, epigenetic heredity has been demonstrated in monocellular eukaryotes such as daphne, in plants or even in humans in the case of predisposition to insulin-resistant diabetes. Niche heredity is also an important factor in transmitting a heritage that is no longer genetic but environmental. The transmission of intestinal microbiota in mammals or insects is also largely hereditary, and its significance in the transmission of pathological characters is not well known. In some species that pay a lot of attention to the youngs, there is a transmission of care that carries a biological effect. Finally, we can observe certain forms of culture which also represent heredity. This is the case of the learning of hunting in lions or singing in birds for example. Heredity is thus carried by a set of mechanisms which contribute to varying degrees to the resemblance between an individual and his offspring. It is proposed to treat all these mechanisms as a single set of hereditary information.[64]

Among all these mechanisms, however, genetic information is considered to be largely predominant and genetics is now considered by most biologists to be the science of heredity. Indeed, the concept of genetic program in principle easily explains the resemblance between relatives: the parents transmit their genome to their children and the program simply reproduces from one generation to the next. It's like copying the same

[63] We refer the reader to the very complete book of Pigliucci, M. and Muller, G.B. Evolution, the extended, synthesis.

[64] Danchin, E. 2013. Trends Ecol. Evol. 2013; 28: 351–8.

computer program to a new computer. Heredity is the domain where the concept of genetic program is triumphant. Pushed to its limits, it leads to the theory of the selfish gene where living organisms are avatars of a genome that reproduces itself eternally, following the genealogy of living beings, from generation to generation.[65] It is known, however, that this idea of an all-powerful genetic program is insufficient to explain the genotype/phenotype relationship. The genetic program is lacking in explaining a certain self-realization of the living being, which, in essence, is based on contingent, non-predictable factors. But in return, the theory of self-organization of the living organism also lacks a heredity of this self-organization. Let us see if, by renewing the genotype/phenotype relationship, it is possible to go further on this question.

[65] Richard Dawkins. The selfish gene. See also Gouyon, P.H., Henry, J.P. and Arnould, J. Les avatars du gène.

9

Reproduction and Phenotypic Canalization

That one body can act upon another, at a distance, in a vacuum, without mediation of anything ... is so absurd to me that I believe that no competent scientist can accept it.

Isaac Newton

Non-sexual reproduction is a common situation in the prokaryotic world and is also found in the eukaryotic world in the case of parthenogenesis or in the case of cuttings and layering in plants. Very often the progeny receives the same genome as the parent (neo-mutations and genetic rearrangements apart). Frequently, it also receives the same environment. The phenotype of the progeny is usually very close to the parent phenotype. In appearance, all this seems therefore simple to explain: to the same factors of causality G and E corresponds the same phenotype P.

However, if we consider that the realization of the phenotype is probabilistic, the mechanism of heredity cannot be as simple. An individual and his descendants share the same genome and therefore have the same probabilities of realizing their phenotype. But this does not imply that their phenotypes are identical. Indeed, the values of the phenotypic traits which can be taken on during the realization of the organism are numerous (Figure 1). Thus, the genotype/phenotype relationship as defined above implies a great phenotypic heterogeneity between the relatives. Indeed, the observation of all possible phenotypes in a large number of offspring should, in principle, provide an estimate of the phenotypic probabilities for the same shared genotype. However, very large phenotypic differences between related parties is generally not what is observed. Most often, the resemblance between an individual and his offspring is close.

We have seen that the environment makes it possible to partially modulate the field of possible phenotypes. Relatives often share their environment in addition to their genome. One might then think that the shared environment explains the excess of resemblance between relatives. However, the environment cannot be the only explanation. The environment brings elements of contingency that play on certain traits (the color of hydrangeas for example). Sometimes it carries information that differentiates the phenotype which is the case of sex in turtles. But it is not clear today that the environment determines the most complex traits. Moreover, the environment is changing from one generation to the next and it seems difficult to explain the phenotypic continuity observed in the living cell lines. This has been discussed for complex diseases or traits. Finally, and above all, the phenotype resulting from the interaction between gene and environment remains probabilistic (Figure 2). It cannot therefore resolve, by nature, the resemblance observed between the relatives.

There must therefore be additional information to transmit the phenotypic resemblance between the relatives which is neither genetic nor environmental. This inheritance must relate not to the genome and environment but to the GxE reaction norm. This deduction is the direct logical consequence of (i) the probabilistic nature of the genotype/phenotype relationship and (ii) the resemblance between the relatives.

If one accepts this conclusion, organisms from the same parent must share with that parent a lasting phenotypic resemblance in time and space (just think of a seed that germinates years and thousands of kilometers away from the mother plant). In other words, there must be additional information that channels the phenotype of the new organism in favour of the ancestral phenotype among possible phenotypes (Figure 3). We will give this phenomenon the name of phenotypic canalization. We can then say that the phenotype is a probabilistic function of the genome, the environment and canalization (noted here C): $P = f(G, E, C)$ where f remains a probabilistic function.

We know that genetic information is essentially structural. It deals with the molecular bricks of the living. We have also seen that, to explain the living, hereditary information enabling its self-organisation was lacking. The information provided by phenotypic canalization must therefore be organizational rather than structural. In other words, it should focus on the dynamics of the biological network rather than on the nature of the nodes.

If we do not want to return to a determinism using a program of a finalist nature, we must admit that phenotypic canalization carries relative information. The most economical hypothesis is that it only reinforces the probability of ancestral phenotypes compared to other possible phenotypes.

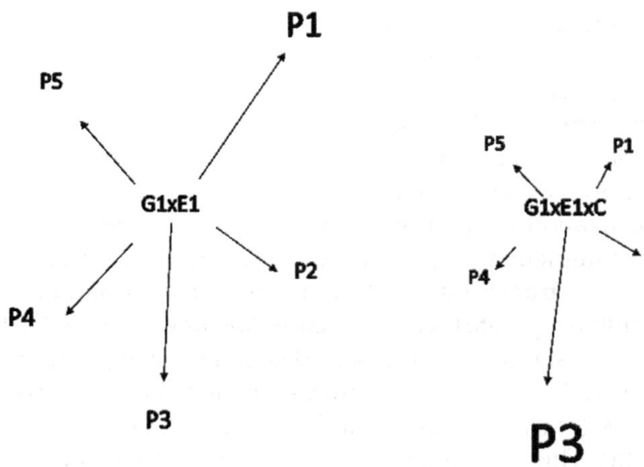

Figure 3. Effect of the phenotypic canalization (C) on the phenotype. Among the possibilities defined by the genome and the environment (left), phenotypic canalization favors the most frequent phenotypes in the ascendants (here P3, right). It therefore greatly increases the probability of certain possible phenotypes.

It is in fact the very definition of heredity. Thus, we will assume, by principle of parsimony, that the phenotypic impact carried by canalization is correlated solely to the match and therefore, in general, to the anterior phenotypic realization in the ascendants. It is transmitted from generation to generation simply by the fact that the organisms are related to each other (Figure 4).

This organizational information must necessarily overlap with genetic and environmental information. It sorts among possible phenotypes derived from the GxE reaction norm. It is a heredity of the GxE reaction norm in favor of the choice already made by the parent. In other words, $P_{i+1} = f(G, E, (GxE)_i)$ where i represents a given generation.

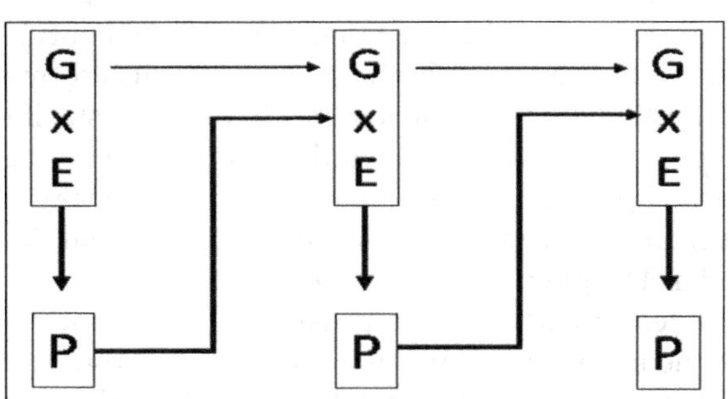

Figure 4. Principle of phenotypic canalization. The phenotype achieved in a previous generation modulates the GxE reaction norm (i.e., the probability of phenotype realization) in the next generation, in favor of the value realized in previous generations.

It must be understood that phenotypic canalization is not (and cannot be) of genetic origin (in the modern sense of the word "gene"). Genetic information, by encoding proteins, provides the biological molecules that are used to construct life. It carries structural information that defines the field of possibilities and brings the building bricks of the house. On the contrary, phenotypic canalization carries organizational information that directs the realization of the organism towards the parental phenotype. It is the way of implementing the house by following a plan available by tradition. Progeny transmission of both types of information, however, makes it difficult to separate them in practice. And as we are used to thinking in terms of genetics, it follows that the idea of a genetic program bearing both structural and organizational information is easily accepted. But this is not obligatory (at least in theory) and it is possible to dissociate the two types of heredity. This is what the idea of phenotypic canalization suggests.

It will be objected that hypothesising organizational heredity superimposed on structural genetic heredity is only an artifice. What change can make a distinction between structural and organizational information if they are both transmitted simultaneously from the ascendant to the offspring? In fact, as we shall see, the distinction between the modes of transmission of this information quite clearly modifies the way in which certain biological questions are envisaged, since it brings phenotypic hereditary information, relating to the matching alone (and not to the genome composition or environmental exposure).

With phenotypic canalization, the individual phenotype of each daughter cell can be seen as conditioned by the phenotype of the parental cell. Throughout the generations, this system has resulted in a group of cells or organisms physically isolated but connected in a system where each object remains largely identical to the others. Thus, a colony of bacteria or aphids forms a set of organisms replicated identically. The same applies to plants obtained by vegetative replication, such as orchids sold in stores.

Due to phenotypic canalization, asexual reproduction is normative. At most, it limits the expression of a genome to a single phenotype: that of the parent. The space of the possible phenotypes is reduced by canalization (even if the definition of the phenotype is still probabilistic if we want the reasoning proposed here to remain valid). It is as if the possible values of the phenotype had probabilities modulated by the existence of the ancestral phenotypes already realized. The variance of intra-familial characters then becomes smaller. The phenotype is canalised.

Due to the lesser extent of the possible range, phenotypic changes only occur in the event of major mutations and environmental changes. Indeed, the phenotype changes only if the field of possibilities defined by the genome no longer contains the canalised phenotype or if its realization becomes impossible (or very unlikely) because of environmental constraints.

In other words, this means that the GxE reaction norm can no longer lead to the channelled phenotype for the latter to change. Consequently, in a constant environment, the notion of a genetic program regains its rights and the genotype/phenotype correlation appears precisely related. The mutation becomes the apparent motor of phenotypic variation. We are in a domain more genetically deterministic of the living, that of traditional genetics with a univocal relation between the genotype and the phenotype.

It must be understood, however, that, according to our hypothesis, this univocal relationship is not the result of a genotype/phenotype relationship that is unambiguous in itself. It is in fact due to the canalization which produces one or a few phenotypes among the field of possibilities defined by the genome. In the absence of a "buffer effect" linked to the canalization, any genome (with or without a neo-mutation) should produce multiple phenotypes and the mutation would have only a relative effect. Mutation is only the apparent cause of a phenotypic change. It has a phenotypic effect because the canalization in favor of the realization of the usual phenotype is impossible due to its presence. We have already discussed this point for Mendelian diseases: a mutation has a reproducible effect only if the field of possible phenotypes is amputated. It is also because of phenotypic canalization that its phenotypic effect is maintained over the generations.

Phenotypic canalization is transmitted with mitosis and can thus also participate in the resemblance between cells of the same lineage in a multicellular organism. It can therefore enhance cell differentiation. A particular case of mitosis is the case of monozygotic twins. These twins come from a single fertilized egg. The organisms thus separated at the time of the first mitoses. If one continues the idea of phenotypic canalization, monozygotic twins are therefore in every way entangled. This may explain their almost perfect resemblance, throughout their life and up to later ages, including for twins raised separately. This almost perfect resemblance of monozygotic twins would be misunderstood with a loose probabilistic phenotype that does not call for a particular resemblance between them. This almost perfect resemblance is, on the other hand, explicable, without going back to a genetic program, in the case of phenotypic canalization.

A particular case of asexual reproduction is that of cloning. In this situation, a de-differentiated cell nucleus is transposed into an enucleated egg. The clone will develop by successive mitoses from this egg, forming a complete organism, almost exactly identical to the organism of origin. In the case of a progeny of mitosis, the phenotypic ducting of the two organisms is expected, explaining their resemblance, which would be difficult to consider under a model that is largely probabilistic. Another close situation is that of cuttings or grafts in plants. In this situation too, cuttings or grafts reproduce identically the parental phenotype, which makes it possible to retain certain unstable (or rather unlikely) phenotypes.

10

Fertilization and Overlapping of Genotypes

Any situation in quantum mechanics can be explained by saying, "Do you remember the experience of the two slots? Well, it's the same thing."

Richard Feynman

In the more complex case of a sexual reproduction, the genotype of the offspring usually consists of a random and equal mixture of the two parental genotypes. This mixture is made during fertilization from two gametes derived from meiosis.[66] Fertilization results in a mixture of two parental hemi-genomes to produce a strictly new genome. For complex organisms, this particular genome never existed and will probably never exist again. Indeed, it is very unlikely that the same genome will emerge again by chance. We are all genetically different from each other. Unique individuals forever.

The fusion of the two parental hemi-genomes cannot be considered as a simple addition of the information transmitted by the parents. The genetic information received from two parents interfere with one another. If we follow the analogy between biology and quantum physics, fertilization can be compared to a superposition of the "wave functions" corresponding to the two parental hemi-genomes. Superposition is a fundamental property in quantum mechanics. It allows wave functions to combine, resulting in

[66] Meiosis is the mechanism of recombination followed by cell division which presides over the formation of gametes.

complex waves. It results in redistributing the probabilities of observing certain states during physical measurements made on the particles. Similarly, fertilization results in redistributing the probabilities of the elementary phenotypes corresponding to the new genome of the child organism. Sexuality is to biology what the addition of wave functions is to quantum physics.

The newly created genome thus carries with it possible new phenotypes. If the environment of the child differs from that of the parents, then we should observe great dissimilarities from one generation to the next. Here again, it is not what we observe and the children usually resemble their parents. We will therefore assume that sexual reproduction also induces phenotypic canalization from an individual to its progeny, making the phenotype of the progeny closer to the parental phenotype than expected from the distribution of probabilities associated with the GxE reaction norm.

However, this canalization cannot be carried out "as a whole", as for asexual reproduction, because it would then be necessary for the descendants to resemble only one of their two parents, which is of course not observed. Thus it is necessary to form a hypothesis of an elementary canalization for an elementary trait. In the case where the genetic information transmitted by each parent is interchangeable and does not interfere with each other, it can be assumed that the probabilities of achieving one or the other elementary parental phenotypic traits are then identical and close to 0.5. For each elementary trait, the phenotype of the offspring is thus determined (almost) one out of two by one parent or identical (almost) to one out of two by the other parent (Figure 5). There is trait-by-trait canalization on parental phenotypes.

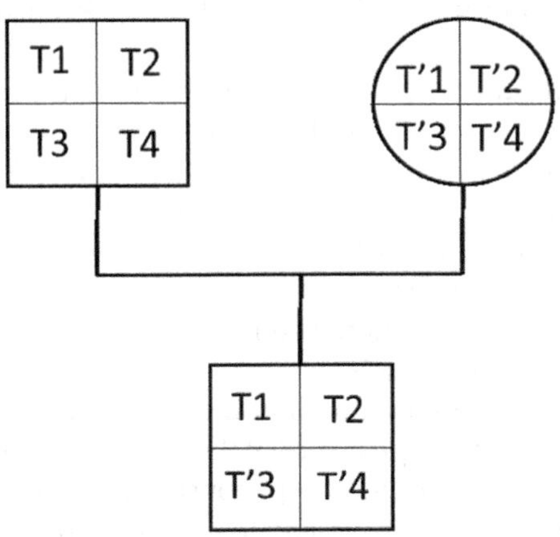

Figure 5. In the case of sexual reproduction and for each trait (T1 to T4) the value of the parent trait or parental trait (T or T') is transmitted to the offspring with a probability close to 0.5. Rarely, a new value of an elementary phenotypic character (T'' for example, not shown here), not expressed in the ancestors, may appear.

In the particular case of Mendelian characters, genetic information can be summarized by the effect of a single genetic polymorphism that conditions the phenotype of a single protein. The idea of canalization restores Mendel's laws which specify that the phenotype of the progeny is predicted by the phenotype of the ascendants and that the characters are transmitted in a discontinuous manner.

Interference between parental genomes is well known for Mendelian traits. Thus, the interference between dominant and recessive alleles redistributes the probabilities of realization of the elementary character. A dominant allele prevents the phenotypic expression of the recessive allele. From a molecular point of view, recessivity is most often associated with loss of gene function. Consequently, a gene carrying a recessive variation in the heterozygous state is not compatible with expression of the recessive character. Indeed, how could this defective character then be realized if the normal gene is expressed? Conversely, a recessive allele encoding a defective protein in the homozygous state results in a complete loss of function which can be expressed in the form of a novel phenotype not expressed in the lineage previously. Each time, the defective character is or is not in the field of possible phenotypes compatible with the genome.

In this example, it will be noted that Mendelian traits do not substitute phenotypic canalization for genetic and environmental causes, but are added to increase the probability of certain phenotypes if these are possible. In the case of dominant/recessive alleles, the role of phenotypic canalization appears to be limited, since genetic determinism is strong (i.e., the field of possibilities for a given character is limited by the mutation carrying the Mendelian effect). As with non-sexual transmission, the mutation has a strong effect if it amputates the range of possible ancestral phenotypic values.

If the application of the hypothesis of phenotypic canalization to Mendelian characters amounts to falling back on known laws, it is more interesting for complex traits where all variations of the genome are supposed to participate more or less in each phenotypic character. In this situation, defining the interactions between parental genomes is much more complex. According to a threshold polygenic approach where parental alleles involved in the elementary phenotypic character are simply added, calculations indicate that in each generation, gene mixing should virtually eliminate all qualitative innovations in the second generation, with little likelihood of seeing the re-appearance of the same trait value in subsequent generations in the same family (except consanguinity). Indeed, the allelic mixture in each generation should lead to a regression towards the average of the value of the phenotypic traits. Their variance should be reduced rapidly in a given genealogy and therefore the phenotypic polymorphism in the line would disappear.

On the contrary, with canalization, the polygenic elementary characters can also be transmitted in a discontinuous manner. The elementary phenotypic character is relatively preserved in the genealogy by the fact that the phenotypic information is maintained in time. Mendel's laws are to some degree extended to polygenic characters. Elementary phenotypic characters, although complex in determinism, can be transmitted from one generation to the other as such, discontinuously, without mixing the parental characters (except interference).

It should be noted that with the hypothesis of phenotypic canalization, the elementary phenotypic character becomes not only a module of the biological network and a measurable physicochemical value but also a unit of hereditary transmission. Phenotypic canalization, as defined here, gives to the elementary trait the meaning of the word "gene" as defined by Johannsen in 1909. This is not surprising because phenotypic canalization merely refocuses heredity on the phenotype. The elementary character is both the unit of phenotype and the unit of transmission by canalization.

In the absence of major interference between parental hemi-genomes, the probability of achievement of each parental elementary trait should be close to 0.5. But in general, it must also be admitted that the probabilistic character of the genotype/phenotype relationship implies the possibility of seeing new phenotypes simply by chance. The determinism carried by the canalization cannot be complete if the phenotype is probabilistic in nature. As in the case of asexual reproduction, there is always a slight probability of a phenotype other than that defined from the parental phenotypes (genetic and environmental factors being identical or not).

What is transmitted from generation to generation by the phenotypic canalization is a greater propensity to carry out a phenotype already realised in the ascendants. It is therefore a memory of what has already been done in the past. But if this memory passes from one generation to the next, is it not possible that it passes through several generations? Let us then follow the transmission over the generations of the phenotype with canalization. In the first generation, the risk of recurrence in the son of an elementary character present in the father is close to 0.5 (in the absence of interference). In the next generation, the phenotype of the son will participate in the definition of that of the grandson but the phenotype of the grandfather will keep a probability of realization lower but not zero. Thus, from generation to generation, the probability of recurrence of a phenotypic trait decreases progressively in the genealogy but never completely ruled out. It decreases with the number of generations where the phenotype is not realized and re-increases with each new realization in the line.

A consequence of phenotypic canalization is that, even for very ancient ancestral phenotypes, there is a low likelihood of resurgence. This phenomenon is known as atavism: a phenotypic character of an ascendant can occur occasionally in a descendant, in a random and unexpected way. It is thus that one finds the nose of the great-grandfather in his great-grandchildren, that zebra markings appear in some horses and that legs grow to some serpents.

From this point of view, phenotypic canalization then leads to an accumulation of possible values for the phenotypic character that are increasingly numerous over generations. But in each generation, the probability of each is updated by the realization of the new phenotype and the oldest values become less and less probable. This way of seeing recalls the integral of the paths in which Feynman schematizes all the possible modalities of occurrence of a microscopic event in quantum physics.[67] The probability of realizing the event is then given by the sum of all these modalities, some of which are very probable, while many have very low probabilities. In an analogous manner, the different phenotypic values can each be considered as a possible phenotypic pathway. An individual is the weighted sum of all his ancestors.[68]

The mode of inheritance associated with sexual reproduction exploits the modularity of the phenotype and allows a renewed combinatorium in each living individual resulting from non-consanguineous sexual reproduction. It thus opposes asexual reproduction which is normative for the global phenotype. On the contrary, sexual reproduction creates new phenotypic combinations in each generation without any new mutations occurring by rearranging different values for the elementary characters. Canalization, however, restrains the creativity resulting from fertilization because it largely retains its normative character for each elementary phenotypic trait. Sexuality has therefore not completely liberated the living from the normative constraint since it remains based on a cell division with phenotypic canalization.

When a novel trait appears as a simple probabilistic effect of producing a genome in contact with a given environment, its frequency is low in the population and its probability of recurrence decreases rapidly over generations. It is therefore unlikely that it will persist over time. Phenotypic canalization favors the most frequent ancestral traits. As a consequence, probabilistic phenotypic characters can only stabilize in the long-term (i) by random recurrence in the population (especially in the case of inbreeding or low reproductive numbers) or (ii) by extreme selection for several generations or (iii) by stable changes of the genome (i.e., mutations) or of the environment affecting the possibilities of achieving the elementary character. Here we retrieve the classic data of population genetics.

---

[67] See Feynman, R. Light and matter.

[68] With sexual reproduction, the ancestors of a given individual are in fact numerous and the pool of phenotypes is largely shared in a panmixic population.

Because of this normative effect, phylogenetically the oldest characters become, over time, more and more stable. It is understandable, then, that the traits which appeared the earliest in the history of life are also the most monomorphic, apart from any consideration of the more or less fundamental nature of a character for the survival of an organism. The molecular symbiosis of protocells is extremely stable (e.g., the genetic code). It is very unlikely that an animal will develop with four embryonic layers instead of three. It is somewhat less unlikely that a snake will find legs or a horse zebra stripes. It is also understood that the early stages of embryogenesis are common to several lines and that the later stages are more diversified.

On the other hand, the most recent phylogenetic characters must also be the most polymorphic since they have a greater variance possible with identical genotype and environment. Thus, the front line of progress of the living is visible in the places of great diversity which correspond to the novelty that is tested. A little as if the line of the living was exploring the ground for new achievements. Just as walking involves a lot of trial and error when it is acquired in an infant and then becomes automatic and unconscious in a few years, phenotypic canalization gradually assimilates the phenotypic characters that become stable gains over generations.[69]

[69] For a discussion on the front of progress see Schrodinger, E. Mind and matter.

11

Nature of Canalization

The discomfort I feel comes from the fact that the perfect quantum correlations that are observed seem to require a sort of "genetic" hypothesis. For me, it is reasonable to assume that the photons in these experiments carry with them programs, which are correlated in advance, and which dictate their behavior. ... [But] what is reasonable simply does not work.

John Bell

With the canalization mechanism, we hypothesize that two related organisms are close; that they are somehow resonant. Everything happens as if there was information that linked the two phenotypes, somewhat like a behavior of mimicry between relatives. The question that immediately arises is then to know what is the support of the information shared between ascendants and descendants. If this support can be brought by epigenetics for asexual reproduction, this does not seem to be the case today for fertilization where the known epigenetic marks are inactivated during meiosis, much like a reset for a computer. The support of the information thus remains hypothetical and this is an important flaw of the hypothesis advanced here. In any case, it is necessary to imagine that this is information of a functional nature such as epigenetic marks or rules of operation for automata (see above). The support is therefore expected to be subtle and in some way super-imposed on traditional genetic information since the phenotypic canalization indicates how this should be used. This is epigenetic information in a broad sense as assumed by Waddington[70] but still to be fully understood.

[70] For an overview of Waddington's work, see Van Speybroeck, L. 2002. Ann. NY Acad. Sci. 981: 61–81.

Phenotypic canalization recalls quantum entanglement. Two particles are said to be entangled if the measurements that can be made on them are correlated (positively or negatively). Thus, although the result of the measurement is probabilistic for each of the particles and for all the two particles, the measurement made on one particle is necessarily correlated with the measurement made on the other. This correlation does not depend on the distance and the time between the two particles and therefore the two measurements. It is not due to any "communication" between particles. The concept of entanglement is strange and difficult to manipulate because it makes use of a link between particles that makes them somehow intrinsically linked although they are physically separate entities. In fact, entanglement forces two associated particles A and B to be seen as an inseparable A + B system. For a given observable, the system A + B cannot have all the possible values as if the particles were independent. If the particles A and B, which can each take either the values + or – for a given observable, are negatively correlated, the system A + B can only take the values +/– and –/+ then the values +/+ and –/– are prohibited. In the same way, for positively correlated particles, the system A + B takes the values +/+ or –/– but never +/– or –/+. An entangled system is therefore a less free system. The whole is less than the sum of the parts. Canalization has an identical effect. Although the phenotypes of the bound organisms are probabilistic, they are positively correlated. In general, related organizations are interdependent in terms of their organization because of their single parentage. This seems particularly explicit for monozygotic twins.

Such an entanglement may seem trivial in biology where the living units are organized in complex networks with feedback between modules and where there are therefore strong correlations between certain modules or phenotypic characters. An organ may be given by way of illustration in which two cell types A and B participate in two opposite regulatory pathways involved in inhibition and the other in the activation of the same metabolic pathway (insulin or glucagon cells of the pancreas which regulate blood glucose levels for example). If A (insulin) is activated, it can be deduced that B (glucagon) is inactivated and vice versa. The +/+ and –/– states of the system A + B are forbidden as in a system entangled negatively. At first sight, this phenomenon can therefore be considered comparable to entanglement in the quantum sense. As they participate in the same biological function, the state of the cells is perfectly correlated and the cells of type A and B form an inseparable whole like two entangled particles. But there is one notable difference: the co-determination of the cells is linked to the simultaneity of the determination of the two protein states (by glycaemia). There is common causality: blood sugar level. On the contrary, quantum entanglement is not due to a common causal relationship. It is not linked to a joint determination. The measurements made on two particles are independent. They depend neither on the moment nor the place where the measurement takes place.

Similarly, phenotypic canalization does not express a common immediate causality. It correlates related phenotypes without assuming immediate causes (genetic and environmental) that determine the phenotype. In the case of a parent-child pair, the phenotype of the child appears to be conditioned to the parental phenotype by the phenotypic canalization, since the latter is carried out later. The link between organisms is asymmetrical because of the birth asynchrony. It therefore appears as a constraint on the offspring. But if we go back to the case of monozygotic twins, the phenotypic canalization must be understood as the fact that their phenotypes are simultaneously determined at each moment by identical "entanglement" with their ancestors and not only because of their shared genome and environment. In a particular case, two organisms sharing the same genome but from different parents should have different phenotypes. However, there is no such situation in the nature of genetically identical and unrelated organisms that could test this idea. However, we may have the answer in the future with synthetic biology.

To explain the notions developed here, animals like hydra also seem to me exemplary. *Nanomia* are made up of "persons" from the same egg and living separated or grouped in colonies. In the case of colony formation, initially isolated individuals differentiate in segments (mouth, tentacle, reproductive tract, etc.) that meet. Each person adapts spontaneously to his collective function by expressing certain genes useful for the global organism.[71] The arrangement of all these "segments" then forms a single animal resembling a jellyfish.[72] Self-organization is not directed by a specific "person" and does not seem predetermined in advance, at least according to our current knowledge. If true, it is a lasting interference between people from the same colony. It is not known what each person will form as a segment but it is known that all people will each form a different segment to result in a coherent animal.

Just as the medium explaining the quantum entanglement remains unknown, the substratum carrying the channeled information remains to be determined. Despite these difficulties, let us try to move on. If one asks about the nature of the information shared between generations, one can say that the genome brings the biological bricks of the living in quality and quantity. It gives information about the nodes of the network. It therefore provides a set of possibilities virtually very large. The environment provides the contextual data that constrains the construction of the organism at the time of implementation. It also modulates the quality of the nodes. The phenotypic canalization provides a form of implementation by proposing

[71] See Siebert, S. et al. 2011. PLOS One, 6: e22953.

[72] See also, S.J. Gould's article on siphonophores. The smile of the pink flamingo, Chapter 5.

a dynamic organization of the nodes between them. It carries a relational information. Thus, canalization provides information not on the nature of the nodes but on the dynamics of the flows that pass through each part of the network.

Let us take an image. If a person walks in the countryside, it describes a journey that brings together, for example, two villages. This path is only slightly visible because the passage of a single person leaves little trace. But if multiple people regularly follow the same route, a pathway is created and then a road. And the following people will increasingly use the existing path. Canalization is a procedure that unites two biological elements just as a route is a path between two points. This procedure is reinforced whenever it is implemented to become the norm. While genetics and the environment contribute to the fabrication of nodes (villages), the canalization indicates how the nodes of the biological network must be connected (the path). It is the memory of the links established between the living units. It shows how to build the biological network from a functional point of view.

To explain the operative side of phenotypic canalization, let us take the example of a house. The building materials available (bricks, wood, cement, glass, tiles, etc.) are analogous to the products of the genes and will condition the realization of the house but the field of possibilities is immense. The environment will condition the realization of the house either by playing on the nature of the materials (nature of the wood, constituting bricks) or by playing on the possibilities of implementation (nature of the ground, rain or cold may influence the final realization). The implementation will provide a project, i.e., a way of doing, generally following the local tradition of the region (a more or less steep roof, orientation of the facades, size of the openings, etc.). The house at the end will be the resultant of the three types of information. Let's take another example: a cooked dish. Its realization depends on the ingredients available, the tools and modes of cooking accessible and finally the recipe and know-how of the cook. One could thus multiply the examples ad infinitum.

The supplementary hereditary information represented by the phenotypic canalization seems to be of the same nature as an operational process, a way of implementing an action. It is a kind of tradition. Under the guise that genetic information and the environment are compatible, it is transmitted through the generations. It is a heredity of the GxE reaction norm. This is in a way the stabilization of the function symbolized by the "x" of the formula "GxE" over generations. Like the house and the recipe, the result of the implementation is predicted at least approximately if the conditions of realization are not changed too much. The phenotypic canalization then ensures the perenniality of the organization of the living biological network, an essential property that is so far lacking to link the concept of genetic program and the concept of self-organized network of living organisms.

The phenotypic canalization can carry only hereditary information since canalization is by definition a phenotypic relation between relatives. Like tradition, its only reason for existence is simply due to its previous existence! It is maintained and maintained by reproduction. It is its corollary. It was able to start with the functioning of the daughter protocells sharing the same molecular networks and gradually refine to become subtle and complex. The information on the phenotype shared by two organisms as well as the entanglement of two quantum particles touches on their matching. There is no canalization without a common history. The phenotypic canalization reinforces the historical nature of the living being. This information can be seen as mutual re-knowledge of organisms of the same lineage. It does not exactly resemble a memory since it is actualized in each generation and does not reproduce a universal archetype. The term tradition seems more appropriate. It implies the reinterpretation, for each new organism, of the transmitted phenotype. It indicates a lively and up-to-date transmission.

One can also think of an imitation or rather an empathy. Empathy is an ability to resonate in the state of another person demonstrated in humans and certain animals (elephants, monkeys, cetaceans, corvids). It implies a recognition of the state of the individual and a reconstruction of that state in his own interior. The empathetic state activates the motor areas corresponding to the observed state (the mirror neurons). It induces, to some extent, a behavior of larval imitation which can be at the basis of the learning of the young person. Empathy is therefore to feel oneself in the same phenotypic state as the other which corresponds well to the idea of the phenotypic canalization between ascendants and descendants.

The link between this operative information and genetic and environmental information is complex. Phenotypic canalization increases the probability of recurrence of a phenotype already achieved. It stabilizes the GxE response norm across generations. It is therefore completely conditioned by the existing phenotypes which are themselves under the dependence of G and E. Phenotypic canalization is not creative. What creates is the interaction between the genome and the environment. The phenotypic canalization is passive. It only transmits a phenotypic information that already exists, just as DNA replication merely reproduces the genetic information that will be distributed between two daughter cells.[73]

On the contrary, for genetic and/or environmental variations of modest amplitudes, the phenotypic canalization maintains a stable state. It is opposed to the diversity of the living being in the same genealogy. Phenotypic canalization goes beyond and encompasses genetic and environmental information. As long as the genome and the environment allow it, the phenotypic canalization produces the traditional phenotype.

[73] The only creative element is actually the living organism which self-realizes itself from the three types of information.

Genetic and environmental information is only permissive. The phenotypic canalization then carries the essential part of heredity. As a direct consequence, most mutations of a complex genome appear neutral.

Phenotypic canalization is conservative. It will behave as a force of inertia for phenotypic changes induced by the genome or the environment. It stabilizes the "traditional" phenotype. On the other hand, if the environment or the genome is changed (in a brutal way or by slow drift), the "traditional" phenotype cannot be maintained. There will then be (rapid or progressive) evolution of the phenotype. The phenotypic canalization will then have a facilitating effect by stabilizing the new phenotype resulting from a neo-mutation or a new environment. It thus has a paradoxically double role, stabilizing the traditional phenotype or the innovative phenotype according to the context.

Let us resume the analogy with the house. Phenotypic canalization, analogous to the tradition of construction, maintains and reinforces phenotypic elements derived from genetics (the use of bricks for example) or environmental (use of earth to join bricks for example). In case of genetic change (the disappearance of bricks in favor of breeze blocks for example), it will maintain the overall structure of the house, adapting the construction. In case of impossibility, it will stabilize a new typology of house adapted to the new constructive constraint. In case of environmental change (the discovery of concrete for example), the canalization will do the same.

We have seen in the chapter on behavior that a biological network representing a living organism was partly defined by its areas of greater variability and greater fixity. The organism cannot adapt to all its parameters at the same time and each species is characterized by points of greater phenotypic fixity that make it possible to recognize it. A man has a temperature set at 37°C but a variable adipose mass. A termite has a particular diet and microbiota but a variable temperature, etc. The phenotypic canalization refers to these fixed characteristics of the biological network. It defines, as it were, the "constants" of the species corresponding to the fixed points of the biological network.

Phenotypic canalization must be seen as a general process that applies to the biological network at all hierarchical levels of the living (molecule, cell, organism, society). The lowest level of the scale should correspond to the protein folding mode. At the cellular level, the canalization must correspond to epigenetic mechanisms (DNA methylation, small non-coding RNA, histone modification). At the level of the organism, the canalization must be concerned with the establishment of development. Finally, at the highest level of the ladder, the canalization must correspond to the transmission of operating procedures to achieve extended phenotypes or the organization of animal societies.

12

Canalization: Experimental Data

The other major evolutionary effect of the epigenetic system is that the development of organisms is partly channeled in the sense that, although it may be altered by the effect of environmental stress, it also shows a tendency to achieving a normal end result despite disruptive circumstances.

Conrad Waddington

We have seen that phenotypic canalization is constraining. It reduces in practice the number of states that a phenotype can take. It channels the realization of the living to a smaller number of possibilities defined from information of genetic and environmental origin and conditioned by the parental phenotype. But, unlike genetic information that relates to the structure of RNAs and proteins and determines living organisms at the most molecular levels, phenotypic canalization exerts its effect at all scales, in particular at the highest scales, corresponding to the forefront of the living. It goes without saying that phenotypic canalization exerts its action through the intermediary of the biological molecules. It therefore has an epigenetic effect (by taking the term in the broad sense). But the object on which the canalization is carried is not the gene but the trait. It is not genocentric but phenocentric.

However, according to the current paradigm, the organism results from a strictly deterministic genome-environment interaction and a heredity of the GxE reaction norm suggested by phenotypic canalization is not proposed. It is therefore necessary to provide arguments, if possible convincing, to revise the current model. It is indeed difficult for biologists (of which I am) to accept this new idea without experimental arguments.

Certain facts or concepts, however, seem to support the idea of phenotypic canalization.

Evidence is provided by Waddington's studies of canalization followed by assimilation.[74] Canalization is seen by Waddington as a buffer effect that reduces phenotypic plasticity in relation to environmental exposure. It simplifies the GxE reaction norm by making the phenotype independent of the environment. Genetic assimilation is "the process by which a phenotypic character, originally produced only in response to an environmental factor, becomes, through a selection process, taken over by the genotype in such a way that it occurs even in the absence of the initial environmental factor".

Waddington subjects pupae of wild drosophila at 40°C for 4 hours. In some flies, the treatment creates a phenotype that does not occur spontaneously at 25°C (reference pupae incubation temperature): drosophila have no posterior transverse vein. Two lines of drosophila are then established by selection and iterative crosses: one responding to thermal shock (i.e., with abnormal wings during heat shock) and the other resistant to this effect. Approximately 2,000 flies are studied in each generation. 23% of the animals initially had the heat-induced phenotype.

- In the responder line, this number increases rapidly with selection from the 5th generation (approximately 60%) and then progressively until the 17th generation where the phenotype then concerns 94% of the animals.
- In the non-responding line, despite the selection, the percentage of responding animals fluctuates from 7 to 35% with an average around 15 to 20% throughout the generations studied.
- From generation 15, drosophila of the susceptible line began to have the phenotype usually induced by heat, whereas they were no longer subjected to thermal shock. The frequency of insects with abnormal wings without thermal induction was 67 to 100% according to the experimental conditions and the lines. This frequency remained relatively stable over time.
- When crossing these spontaneously abnormal drosophila with wild drosophila, the abnormal phenotype disappeared rapidly. The crossbreeding of these (back to normal) drosophila with their parents with spontaneous altered wings (backcross) restored the spontaneous phenotype "without transverse vein".

This Waddington experiment shows that a new phenotypic character, induced by an environmental factor, can become stable throughout the generations. By selecting the new phenotype, it becomes more and more frequent in the lineage and ends up being the main phenotype. Moreover,

[74] Waddington, C.H. 1953. Evolution, 7: 118–126.

the cross-breeding experiments with ascendants with the old (respectively new) phenotype show that the new phenotype disappears (respectively reappears) in a manner correlated with the frequency of the phenotype in the ascendants. The new phenotype is revealed by the initial environmental factor but this is not necessary beyond this, since the new phenotype persists in the event of the disappearance of the triggering factor in a few generations. This has only served to create the new phenotype but is not necessary for its transmission over generations after its stabilization in the lineage. We see therefore that the experiment is in complete agreement with what is expected under the hypothesis of phenotypic canalization.

Waddington spoke of canalization followed by genetic assimilation of the character with the generations. The canalization evidences the property of the phenotype to be transmitted (even in the absence of the initiating factor). The notion of genetic assimilation indicates that the stability of the phenotype is reported by Waddington to artificial selection which increases the frequency of the alleles responsible for the new phenotype. It is these genes, indirectly selected by the phenotype selection, that would be responsible for the stabilization of the phenotype despite the loss of environmental exposure. I have preserved the same term of canalization in this work to indicate that I am indeed speaking of the same phenomenon as Waddington. The term "phenotypic" I add, however, indicates that my interpretation is not based on genetic inheritance but on a heredity of the phenotype, contrary to Waddington's interpretation. It is the selection of the phenotype and not the indirect selection of the genotype that explains the canalization.

The work published by Suzanne Rutherford and Susan Lindquist[75] gives similar results to those of Waddington for a new phenotype revealed by a variation, this time either environmental or genetic. The authors show that mutated drosophila for the gene encoding the chaperone protein Heat Shock Protein 90 (Hsp90) have malformations. These anomalies are inconstant and different from one fly line to another. They affect the wings and the eyes. These observations are reproduced by a chemical inhibitor of Hsp90, indicating that it is the functional defect of Hsp90 which is important for inducing the mutant phenotype and that this defect can be of genetic or environmental origin indifferently. As in the Waddington experiment, drosophila mutated on Hsp90 with a malformation of the wings (respectively the eyes) were then selected over several generations by iterative crosses. Flies without malformation were also crossed. The rate of malformations remained low and stable in this line. On the other hand, in the malformed drosophila lines, the rate increased to stabilize from the fifth generation between 60% and 90% depending on the lineage. The

[75] Rutherford, S. et al. 1998. Nature, 396: 336–42.

backcross with the wild-type strain eliminated the phenotype whether it was dependent on genetics or the environment. *De novo* mutations added to the Hsp90 mutation or environmental variations are highly unlikely given the low number of flies studied and the controlled nature of the experiment. More importantly, the retrospective genotyping of the drosophila lines expressing the morbid phenotype shows that they no longer carried the Hsp90 mutation. It is therefore necessary to conclude that the phenotype has been transmitted without its initial genetic cause being transmitted.

In both cases, the character was assimilated whatever the genetic or environmental origin of the phenotypic alteration. It is as if the environmental or genetic nature of the stimulus did not matter. Environment and genetics have an interchangeable role. Character has become hereditary without its initial cause being transmitted. It is the GxE reaction norm (i.e., the phenotype) that is the object of the transmission. This occurred in about 5 generations (corresponding to more than 95% of ancestral morbid phenotypes). The new character (wing or malformed eye) is specific to each line. There is no new phenotype as would be expected with new mutations created by the environmental or genetic factor. Here again, all the data are in complete agreement with the idea of phenotypic canalization.

The current orthodox explanation of this phenomenon of assimilation, however, is quite different. Cryptic mutations (i.e., not having the ability to express phenotypically) are hidden in the genome of normal drosophila. We have seen that these mutations considered to be phenotypically neutral are indeed frequent in eukaryotes.[76] In the normal state, Hsp90 prevents the phenotypic expression of these mutations. The genetic or environmental inhibition of Hsp90, on the contrary, makes it possible to update their previously "buffered" phenotypic effect. The selection of animals carrying the new phenotype enriches the fly population with these cryptic mutations. Due to the high frequency of the at-risk alleles, the phenotype becomes more likely, even in the absence of the inaugural factor, by genetic assimilation of the phenotype.

In his experiments, Waddington also subjected drosophila to ether vapor during their development, revealing bithorax-like malformations which are now known to be due to mutations in the homeotic Ubx gene. Similarly, Waddington observed an assimilation of the trait induced by ether vapor: malformed flies appeared spontaneously, in the absence of exposure to the toxic, after a few generations in the line of flies sensitive to ether. This experiment was taken up by Gibson and Hogness[77] who showed that assimilation was accompanied by an increase in allelic frequencies of polymorphisms located in the Ubx region in the population of malformed flies. This genetic region accounted for 65% to 75% of the calculated trait

[76] Kimura, M. Théorie neutraliste de l'évolution.

[77] Gibson, G. and Hogness, D.S. 1996. Science, 271: 200–203.

heritability. They also observed a decrease in Ubx gene expression. As expected, it was indeed the Ubx gene that carried the bulk of the biological effect.

At first sight, these data confirm the standard interpretation by selection of risk alleles in the assimilation phenomenon. The ether vapors modified the GxE reaction norm in flies carrying Ubx mutations in their genetic background, thus revealing an inapparent phenotypic effect hitherto. Artificial selection increased the frequency of risk alleles of the Ubx gene in breeding stock, which is expected since these alleles carry the biological effect. But there is a problem. Ubx variants carrying the phenotypic effect were present from the beginning of the experiment but were not revealed without exposure to ether vapors. So why can they do it then, when the environmental factor is removed and back to baseline? According to the orthodox interpretation, this is due to the accumulation of various mutations with artificial selection. But it is difficult to call upon possible additional genetic polymorphisms since Ubx carries the main part of the heritability. It is therefore not the accumulation of multiple genetic variants with the selection that allows the assimilation of the trait as proposed in the orthodox interpretation.

This "orthodox" interpretation of assimilation is also called into question by new experiments.[78] Drosophila bearing the "irregular facet" (If) allele of the Kruppel gene express an abnormal ocular phenotype. Expression of the mutated phenotype is promoted by loss of function of other genes such as chaperones and genes involved in embryogenesis and methylation of DNA such as Hsp proteins and the homeotic TrxG gene. Drosophila carrying the If allele and a mutation of the TrxG gene and thus expressing the abnormal ocular phenotype were crossed with Drosophila carrying the only If allele (without TrxG mutation). In spite of the loss of the mutation of TrxG (present only in the first generation), the abnormal phenotype became more and more frequent (75% in generation 5). So there has been canalization and assimilation here too. In this experiment, we again find the problem raised by the experiment with ether vapor. We start from a well defined mutation of the eye. It carries a characteristic phenotypic effect (and not unknown putative cryptic variants). The phenotypic effect is revealed by the transient inhibition of TrxG. The phenotype then becomes the main phenotype while we have returned to the initial conditions (as soon as the second generation) and that we lost the genetic factor that revealed the phenotype carried by If. Here, it has not been possible in one generation to select multiple additional cryptic alleles which confirms that they are not involved. The genetic factor If is necessary and sufficient for the phenotype to be expressed, but it has taken a step back from the normal phenotype to the alternative phenotype. This boost served as a revealer, a

[78] Sollars, V. et al. 2003. Nature Genet, 33: 70–4.

spark, by raising the canalization in favor of the classic phenotype. Once the boost is given, the alternative phenotype is established with no obvious cause other than the appearance of the phenotype itself.

In another experiment, drosophila originating from a consanguineous (thus very little polymorphic) line carrying the *If* allele were subjected to a chemical inhibitor of Hsp90 in order to reveal the abnormal ocular phenotype. Exposure to the chemical agent occurred only in the first generation. Drosophila were then crossed for 13 generations with selection on the ocular phenotype. Approximately 65% of the Drosophila had an abnormal eye from generation 6 to generation 13. Here, too, there is a canalization revealed by a point-in-time environmental phenomenon. This phenotype was partially reversible by chemical agents modifying the acetylation of histones. These experiments confirm the phenomenon of canalization followed by assimilation in another model with an environmental starting point. They suggest that it is not related to the selection of multiple cryptic alleles since the inbred line is very poorly polymorphic and that the effect occurs as early as the second generation, thus without sustainable genetic selection. Finally, they show that histone acetylation, chaperone proteins and homeotic genes play a role in the phenotype.

In Arabidopsis, a plant model, two Hsp90 inhibitors have an identical effect.[79] They produce a wide range of developmental abnormalities, often marked and specific to each plant line. They usually only affect one character of the plant and are not associated with a growth anomaly (on the contrary, the plants are quite vigorous). They occur at a high frequency that depends on the dose of the inhibitor. Anomalies are not specific to a strain but their frequencies vary according to the strains suggesting a genetic background effect. Congenic plant lines confirm that some regions of the genome are statistically associated with certain phenotypes in treated plants. Inhibition of Hsp90 leads to more frequent changes if genetic variability is increased by crossing different lines. Finally, the inhibition of Hsp90 induces multiple modifications of the GxE reaction norm for multiple parameters according to the strains. In Arabidopsis, therefore, results comparable to those obtained in the vinegar fly are observed.

In zebrafish, the genetic or environmental inhibition of Hsp90 induces variable malformations from one animal to another.[80] Inhibition of Hsp90 revealed new ocular malformations in a specific mutant not usually having an ocular lesion. It modulates (more or less) the morbid phenotype associated with monogenic diseases but may also remain neutral with respect to the morbid phenotype even if the mutated protein is a Hsp90 client. It is therefore difficult to establish simple rules.

---

[79] Queitsch, C. 2002 et al. Nature, 417: 618–24.

[80] Yeyati, P.L. et al. 2007. PLoS Genet, 3: e43.

Overall, these canalization and assimilation experiments are in complete agreement with phenotypic canalization. In fact, one cannot imagine more favourable results! The complex phenotypic character is initially revealed by a genetic, environmental or even stochastic[81] factor. It is then retained by artificial selection by bringing the new phenotype to a higher frequency. Over a few generations, it becomes the most frequent and no longer requires the initial factor to be present to express itself. As expected, if a person with the new phenotype is crossed with an individual whose ancestors have never had the phenotype, the recurrence of the trait is reduced (as a majority of the ancestral phenotypes are then wild-type) whereas crosses with an individual whose ancestors have had the morbid phenotype (we then have a majority of morbid ancestral phenotypes), on the contrary, we reinforce the probability of the morbid phenotype.

The facts are therefore compatible with the hypothesis of phenotypic canalization as proposed above. But our interpretation implies an additional and uncommitted strong hypothesis (non-genetic hereditary information) in contrast to the orthodox hypothesis that rests only on genetic heredity. By principle of parsimony, the phenotypic canalization must therefore be rejected if we cannot decide between the two explanations. However, it is not easy to identify arguments in favor of a phenotypic inheritance superimposed to genetic inheritance. In the above-mentioned experiments, the selection of the strains involved both genetic factors (selection of risk genotypes) and possible phenotypic canalization (morbid phenotypes are selected). Indeed, selection on the phenotype eliminates the alleles that are less compatible with the phenotype and therefore enriches in at-risk alleles. But by selecting on the phenotype, we also reinforce the frequency of the morbid phenotype and therefore its probability of recurrence according to the hypothetical phenotypic canalization. It is then impossible to decide in favor of one or the other explanatory hypothesis by the selection experiments. Above all, the two hypotheses are not mutually exclusive but complementary. Thus, the cryptic variations present in the genome are actors of phenotypic plasticity by increasing the number of possible phenotypes. They are therefore useful when a new phenotype appears. The canalization does not create, it only retains what has already been realized from a set of potential phenotypes. Phenotypic canalization is therefore not an alternative hypothesis to the conventional explanation but an additional explanation.

The problem of the nature of the genes involved in assimilation experiments adds to the complexity of the answer to the question posed. The genes involved in the pipeline are not "ordinary" structural genes. The Hsp90 protein is a chaperone protein. It helps to fold unstable proteins in the cell and thus brings to its protein 'clients' the three-dimensional conformation

[81] Milton, C.C. et al. 2006. PLOS One, 1: e75.

necessary for their function.[82] In the event of failure of adequate folding, it directs the poorly folded proteins towards pathways of degradation.[83] Its role is especially important in the case of stress, especially in the event of heat shock, where many proteins are then poorly conformed. Its "client" proteins are specific but numerous and Hsp90 can be considered to have very high connectivity in the biological network. It has been proposed that the normal Hsp90 chaperone protein, by maintaining stable conformation to malformed proteins or by directing them towards deterioration, allows the organism to tolerate the presence of cryptic mutations in the genome. It would protect the body from multiple mutant proteins by ensuring quality control of newly translated proteins. Malformed proteins encoded by cryptic mutants are either "rectified" or eliminated if they occur and the effect of the mutations is consequently buffered. The mutation of Hsp90 (or the inhibition of the protein by a chemical agent) lifts this effect, revealing the functional impact of these usually silent cryptic mutations. Hsp90 was therefore considered as a "capacitor" to reveal or cancel the effect of genetic variations depending on whether the organism is in a normal environment or subjected to thermal stress.[84] In this way, Hsp90 is seen as a way to stabilize a phenotype for a prolonged period and reveal silent mutations in times of stress.[85] It would then play an important role in the evolving capacities of the organizations. It modifies the GxE reaction norm by allowing the appearance of new phenotypes in case of environmental stress.

The phenotypic canalization property is also to stabilize the phenotype. Hsp90, by stabilizing the protein phenotype, then appears as an excellent candidate to bring the biological effect related to the phenomenon of phenotypic canalization on the scale of the protein. It can then be proposed that the genetic or environmental inhibition of Hsp90 lifts the channeling process at the protein level, thus limiting the maintenance of the traditional phenotype. New phenotypes compatible with the genome of individuals affected by the inhibition of Hsp90 appear. These new phenotypes are new realizations of the genome where cryptic mutations, effects of the environment or simply chance occur.

But then, if the canalization uses the chaperone proteins to exert its effect, how can we distinguish the specific effect of the phenotypic channel from the conventional model? We see that this question is difficult and that the arguments to distinguish the genetic orthodox hypothesis from the hypothesis of phenotypic canalization must of necessity be subtle (but to be honest, we did not expect anything obvious!).

---

[82] Saibil, H. 2013. Nature Rev. Mol. Cell Biol., 14: 630–42.

[83] Höhfeld, J. 2001. EMBO reports, 2: 885–90.

[84] Rutherford, S.L. and Lidquist, S. 1998. Nature, 396: 336–42.

[85] Jarosz, D.F. and Lindquist, S. 2010. Science, 310: 1820–4.

However, the predictions from the two models must not be exactly the same. According to the orthodox hypothesis, in the experiments reported above, the chaperones stabilize the proteins encoded by the "mutated" cryptic variants in the case of a traditional phenotype, but they become incapable of doing so for these same variants after the new alternative phenotype. This has been mentioned with experience revealing the effect of cryptic variants of Ubx or ocular malformations revealed by the inhibition of TrxG. The "buffer" molecules have only transiently changed and the initial and final conditions are identical with respect to these genes. So how does a previously repressed molecular phenotype suddenly become accepted? It can be argued that the absence of the "normal" variant of the gene associated with the malformation gives way and reduces competitive access to the chaperone protein for Hsp90. But then, in this case, it must be the "normal" protein and not the "mutated" protein which is chaperonned in the case of a traditional phenotype. Now, in the present explanation, it is the cryptic variants that seem to have to be taken over by the chaperone, which eliminates or conforms them. And if "normal" is synonymous with "chaperonned", it leads us to a canalization of the "normal" phenotype mediated by the chaperones, which then corresponds to the model predictions of phenotypic canalization via Hsp90.

The second observation is that it is often enough for one or several generations who have undergone Hsp90 inhibitory treatment for the new phenotype to stabilize. This observation argues against the selection of additional genetic variants, which would require several generations. On the contrary, it is very much in favor of phenotypic canalization. Indeed, in order for the new phenotype to stabilize in the line, the canalization must reappear rapidly, otherwise the phenotypic innovation will be lost. Canalization makes it possible to stabilize the novelty that has appeared in the line and a loss of a durable canalization would be opposed to this effect.

The third observation is that it is expected that the Hsp90 deficiency leads to various anomalies in the two explanatory models. In the case of the phenotypic canalization, however, an effect is expected on only one or at most a few characters because the phenotypic canalization is carried out trait by trait. On the contrary, in the orthodox "genetic" hypothesis, the loss of inhibition of Hsp90 should concern multiple traits in the same animal since Hsp90 is capable of buffering multiple cryptic alleles of multiple genes. Moreover, according to the same hypothesis, new traits should appear during the experiments of selective crosses, by simple effect of random associations of cryptic variants during the crosses. This is not what seems to be observed. The phenotypic canalization is specific to each line, without the appearance of a firework of anomalies in an animal or in a lineage. This is confirmed by other experiments in Drosophila.[86] This observation is also in favor of phenotypic canalization.

The fourth difference is that in the case of phenotypic canalization, the pool of available cryptic variants should be less critical than in the case of a "genetic" hypothesis to bring about new phenotypes. Indeed, according to our vision of the genotype/phenotype relationship, new phenotypes may appear with a reduced gene pool since the phenotypes are probabilistic. On the contrary, in the classic theory, genetic determinism is strong and involves numerous cryptic alleles for new phenotypes to appear. The available evidence is more in favor of the first interpretation. Indeed, the pool of cryptic alleles appears to be modest in *Arabidopsis* and *Drosophila*. On the contrary, zebrafish whose lines are less consanguineous does not seem to show a wider range of anomalies. This is confirmed by experiments carried out on highly inbred drosophila lines.[87]

The fifth point concerns the selection of drosophila lines insensitive to heat shock in the Waddington and Lindquist experiments. According to the orthodox hypothesis, this selection should gradually remove the cryptic alleles responsible for the phenotype and the lines should become less and less sensitive to the effect of thermal shock. However, it is not the case. These lines maintain a constant rate of malformation of the wings (respectively eyes) over the generations when the stressor is applied. This goes against the "genetic" hypothesis. On the contrary, if one considers that the thermal shock lifts the effect of phenotypic canalization of the traditional phenotype, the recurrence of the trait remains possible since the latter is not determined by the cryptic mutations alone. It is enough that it is part of the field of possibilities of the GxE reaction norm even without cryptic mutations. Here again the difference between the two explanatory hypotheses is that canalization works on phenotypes and that identical genotypes can express different phenotypes in the same environment.

All in all, some arguments, although subtle and speculative, may suggest that phenotypic canalization is plausible. Of course, the idea of phenotypic canalization is disturbing with respect to the current paradigm of heredity, but if one looks at the data without an *a priori* in detail, it seems at least as defensible as the conventional explanation which has been little argued in reality. In all cases, we can see, through the examples presented and the discussions above, that it is possible to test the hypothesis of phenotypic canalization experimentally, making this idea a scientific hypothesis in its own right.

We must add here a word about the mechanism of action of the phenotypic canalization. As we have said, the mode of hereditary transmission of self-organizing information (called phenotypic canalization)

---

86 Milton, C.C. et al. 2006. PLOS One, 1: e75.

87 *ibid.*

remains very speculative. The transmission of organizational information, however, involves biological actors. Its mode of action must therefore pass through genes and proteins which themselves have an evolutionary history and defined functions. One can then expect that the genes currently known to participate in the organization of living organisms are the tools of phenotypic canalization.

Phenotypic canalization is hereditary information that concerns the organization of the biological network on a living organism. Its effect must be applied to each biological unit of the hierarchical scale of the living organism by actors who facilitate the organization of the phenotype. For proteins, it is a matter of organizing their folding or they would take a considerable time before finding the right conformation with large fluctuations from one molecule to another. For the cell, it is a matter of organizing its differentiation. For the multicellular organism, it is a question of organizing its ontogenesis. The fact that assimilation experiments lead back to the genes of chaperone proteins, genes of epigenetic regulation, homeotic genes or micro-RNA[88] is logical. And this is a strong argument in favor of the hypothesis.

The difference between phenotypic canalization model and the conventional model is that in the first case we consider the organizing genes as a proximal cause and not as an ultimate cause. It is the phenotypic canalization that exploits them. They are its tools. The gene is used for the organization function but it is not the organizer itself. Indeed, according to the theory developed here, if the products of the genes named above are the actors of the phenotypic canalization, the information associated with the canalization is not itself genetic. Thus, the Hsp90 chaperone helps to fold proteins. This function is defined by its sequence and is essentially dependent on genetics. But knowing which protein it folds preferentially is a different piece of information which, it is not necessarily genetic (and which seems currently misunderstood). The same can be said of homeotic genes and methylation genes. To say that the phenotypic canalization uses genes to exert its effect does not therefore solve the nature of the medium carrying the information. The question of what defines the privileged target on which the organizing proteins act and therefore the nature of the information transmitted through the generations remains unresolved.

There is sometimes some confusion about the terms used by the authors.[89] The canalization defined by Waddington as a privileged orientation towards a phenotype in the progeny, is a term that can be considered equivalent to the phenotypic canalization discussed here. This is why I used the same expression, adding the phenotypic term to indicate that heredity is centred on the phenotype and not on the genotype. Assimilation

[88] Horsnstein, E. et al. 2006. Nature Genet, 38: S20–S24.

[89] See Crispo, E. 2007. Evolution, 61: 2469–2479.

refers to the fact that the phenotype is carried by genetics over generations and no longer requires the environmental trigger. Assimilation is a direct consequence of canalization and reflects its selective effect on the genome. We will come back to this. MJ West-Ebrehard proposed the term of genetic accommodation instead of the term canalization or assimilation.[90]

Canalization followed by assimilation is a subject that remains controversial in its interpretation but is increasingly studied.[91] We have seen that the phenomenon has been observed in yeast, worms, plants, or fish in the laboratory.[92] It also appears to be observable in nature and to participate in evolutionary mechanisms.[93] The canalization followed by assimilation has thus been proposed as explanatory for a high proportion of phenotypes where there is a morphological asymmetry (e.g., right-left asymmetry for crab or lobster clamps or asymmetry of shells of certain molluscs).[94] It also explains the transition from a sexual determinism of origin initially environmental to a currently genetic determinism in reptiles. It would be involved in the symbiosis between Acacia and ants,[95] in menstruation in women[96] ... Thus, the fact of canalization followed by assimilation is not questionable even if its interpretation can be debated.

[90] West-Ebrehard. Developmental Plasticity and Evolution.

[91] See Pigliucci, M. et al. 2006. J. Exp. Biol., 209: 2362–2367. Pigliucci, M. and Muller, G.B. Evolution the extended synthesis, p. 355. West-Eberhard, M.J. Developmental plasticity and evolution, p. 150. Braendle, C. and Flatt, T. 2006. BioEssays, 28: 868–873.

[92] Jarosz, D.F. and Lindquist, S. 2010. Science, 310: 1820–4. Sangster, T.A. et al. 2007. Plos One, 2: e648.

[93] Hiyama, A. et al. Frontiers in Genetics. 2012, volume 3, article 15. Aubret, F. et al. 2009. Current Biology, 19: 1932–1936. Suzuki, Y. et al. 2006. Science, 311: 650–2.

[94] Palmer, A.R. 2004. Science, 306: 828–33.

[95] Janzen, F.J. and Paukstis, G.L. 1991. Evolution, 45: 435. Heil, M. et al. 2004. Nature, 430: 205.

[96] Emera, D. et al. 2012. Bioessays, 34: 26–35.

13

Implications of Canalization

We must remember that what we observe is not Nature itself, but Nature subjected to our method of questioning.

Werner Heisenberg

If the experiments described above support phenotypic canalization (without definitively demonstrating it), other biological observations that are difficult to understand may find some form of explanation if we accept this hypothesis. Of course, this is not evidence, but only testable arguments or predictions for debate.

Missing Heritability

An obvious consequence of the heredity mode discussed here is that it adds a high proportion of resemblance between relatives that is not borne by the genome. This is an excellent way of explaining the missing heritability observed for many complex genetic traits (see introduction). Heritability (H^2) corresponds to the share of total phenotypic variance (V_P) explained by heredity, usually assimilated to genetics ($H^2 = V_G/V_P$). By redefining heritability as not only linked to the genetic part but also to the phenomenon of canalization, it increases accordingly. It should be noted that this increase is expected to be considerable because the canalization probably contributes much more to the similarity between relatives than the genome alone for a large number of complex phenotypic characters. Such an increase is certainly capable of explaining the missing heritability for multifactorial phenotypic traits.

The argument developed here that phenotypic canalization can explain some of the missing heritability is of course not a proof of its existence. Drosophila experiments, however, suggest that this is indeed the case.[97] Thus, the chemical inhibition of Hsp90 which inhibits the canalization phenomenon completely abolishes the heritability of certain phenotypic traits. Additional studies should be able to answer this question more generally.

Some authors have addressed the issue of the link between complex diseases and canalization. They suggested that genes capable of buffering environmental effects (i.e., those involved in canalization) would define a state of 'biological robustness' and thus protect against the emergence of complex diseases.[98] On the contrary, less effective buffer genes would make them more susceptible to complex diseases. This way of thinking implies, however, that we regularly find the genes involved in the canalization (such as Hsp90) among the genes that predispose to complex genetic diseases, which does not seem to be the case. So complex diseases do not, in my opinion, translate into a deficient canalization.

On the contrary, if we can dissect a complex character (such as a disease) in elementary characters, it is easy to calculate the family recurrence in the relatives by the phenotypic canalization principle. As an example, we mentioned that Crohn's disease could be modeled by the functional state of a dozen modules (or elementary characters) of the biological network. By applying the transmission rule of one of the two elementary parental phenotypes with a probability of 0.95 (to satisfy the probabilistic criterion of the nature of the phenotype realization) and an identical probability of recurrence of the phenotype of one or other parent (i.e., 0.475 being the probability of observing the elementary character of one or the other parent in the offspring), the recurrence of the disease in families is explained fairly well according to the degree of match with the index case (J.M. Victor and J.P. Hugot, unpublished data).

It should be noted that this calculation does not take into account the transmission of risk alleles or environmental factors. It takes into account only ancestral phenotypes. It is therefore inclusive. This is expected since canalization is a function applying to the GxE reaction norm (which already includes genetic and environmental information).

Embryogenesis

The most ancestral characters (i.e., those already present in the older ancestors) usually occur at the beginning of ontogenesis and are present in all individuals of the same higher taxon. This is so for the functioning

---

97 Milton, C.C. et al. 2006. PLOS One, 1: e75.

98 Queitsch, C. et al. 2012. PLOS Genetics, 8: e1003041.

of the cells, the division of the egg, the presence of three embryonic layers, and so on. These ancestral traits are thus ancient, shared by a large number of organisms and fundamental in the organization of the living. It also happens that they are the most monomorphic. They are fixed. An example is the asymmetry of the heart.[99] In all vertebrates the heart is left and grows asymmetrically. This character may occasionally be absent in some individuals. The frequency of this anomaly is 5% in fish, 1 to 2% in amphibians and birds and 0.1% in mammals (0.01% in humans). There is therefore an increasingly marked fixation with phylogenetic evolution. However, the fixation over time of these characters has no reason to be if they are only dependent on genetics, the environment or stochastic factors. Indeed, genetic mutations occurring at random, they should affect as much ancestral characters as recent characters. The same applies to the effect of the environment or random factors.

In order to explain the fixation of ancestral traits over time, they are sometimes called upon to play a fundamental role in the organization of life (i.e., the lethality of possible mutations) and thus to the limits of the plasticity of living beings. However, possible limitations to the plasticity of the living seem contradicted by the facts. Thus, for example, mammals, although recent organisms, have been able to diversify into a very large number of very varied phenotypes.[100] There are 5,162 genera of mammals grouped into 425 families, of which 4,079 genera (300 families) are fossil. They emerged 220 million years ago and diversified mainly from the Cretaceous/Tertiary crisis 65 million years ago.[101] Mammals are all viviparous vertebrates with hair, udder and warm blood, but they are as varied as the hedgehog, the lamentin, the kangaroo, the bat, the rorcal, the shrew, the man, the cow, the tiger, etc. They do not seem to have lost their capacity for innovation by fixing characters that would be particularly fundamental. On the contrary, the phenotypic canalization explains very naturally the monomorphic character of the oldest phenotypic traits, without making any additional hypothesis as to the more or less fundamental nature of certain characteristics. They are more monomorphic only because they have been channeled for a longer time.

The study of organisms with well-formed but poorly positioned structures (called homeotic structures) was the basis of a particularly vivid branch of modern biology. Many homeotic genes have been discovered, explaining in large part how the "genetic program" can build an organism. There are thus genes or combinations of genes which, expressed or

---

[99] Palmer, A.R. 2004. Science, 306: 828–33.

[100] For more details see the beautiful book by Lecointre, G. and Le Guyader, H. Classification phylogénétique du vivant, p. 368.

[101] Data from Hartenberger, J.L. Une brève histoire des mammifères, Chapter 4.

repressed, will lead to the formation of an organ. The best known example is that of the mutant antennapedia of *Drosophila* which has legs instead of antennae on the head. It is then invoked to explain the canalization of the oldest characters that these architect genes serve as a developmental genetic program and that they cannot be mutated without incurring too many alterations of the phenotype.

However, on closer inspection, all of these genes appear more like a toolbox than a fixed program.[102] Thus, the homeotic genes of the *mouse* involved in the formation of the eye can also induce an eye if they are expressed in *Drosophila* who nevertheless develops a completely different facet eye than that of the *mouse*. Similarly, the messenger RNAs of Sog and Chordin which are genes involved in the dorso-ventral determination of the embryo have opposite roles in *Drosophila* and *Xenopus*. Hox genes do not have the same roles in shrimp and fly. Homeotic genes of amphioxius are capable of being expressed specifically in neural crests of mice or chickens, whereas amphioxus does not possess such embryonic structures. The tissue graft taken from the posterior limb of the python (yet involved) near the wing bud of a chicken is capable of generating a wing. The zebrafish has tapes of homoeotic Hox genes different from those of its close relatives like striped bass. It is nonetheless a fish that can be classified in the same taxon.

Through all these examples (which could be multiplied) we see that the genes involved in early embryogenesis are more tools for self-organization than pre-established programs. These observations correspond fairly well to the idea of phenotypic canalization. The living organism uses the genetic means at its disposal to carry out (whenever possible) the canalised phenotype and the homeotic genes are pre-elaborated and practical tools for the realization of developmental modules which are then used for embryogenesis. To recapitulate the analogy of building a home, homeotic genes provide prefabricated building materials such as windows or gypsum wall partitions that help in the construction of the house but do not fundamentally change the construction process which depends on the know-how (in other words, the phenotypic canalization). They are followers and not engines of evolution as the West-Eberhard note.[103]

Permanence of the Phenotype During Life

We have seen in the preceding chapters that the phenotype of an individual could theoretically be different from one moment to another of his life. This is of course not observed (except in toads that can turn into charming princes!). Mechanisms of resilience or buffer effect on the biological network

[102] Examples mentioned here provide from Lambert, D. and Rezsöhazy. Comment des pattes naissent aux serpents, Chapters 7 and 8.

[103] West-Eberhard, M.J. Developmental Pasticity and Evolution, p. 157.

of living organisms have been reported. However, these are mechanisms but what defines the baseline state? What makes a particular state of a biological system remain referent throughout life (or at least for long periods)? While all the molecules and all the cells of my face have been replaced, I am recognized on a photograph from one year to the next. This question may seem simplistic, but to think about it, it is essential. It does not, to my knowledge, have a clear answer.

There is no reason why our face, fingerprints or the shape of our liver should be perennial. The same applies to the shape of the leaves of trees. However, these parameters are not immediately deductible from genetic information or from an environmental or structural constraint. It must be actively maintained by a kind of memory.

By nature, the phenotypic canalization must be exercised at every moment of life. It therefore has a role in maintaining the organization of life. It applies permanent and sustainable organizational information to the biological network. Thus, if the phenotypic canalization exists, it is certainly a way of understanding the difficult question of the permanence of the organization of living beings throughout life. It constrains the network, defines its points of fixity and gives overall stability to the phenotype that is necessary for the essential property of the living, i.e., its phenotypic continuity. It serves as a memory for the organism, defining a state of reference specific to the biological network.

Probably phenotypic canalization also helps to solve internal or external problems and to buffer errors in ontogeny or repair a damaged organism. Thus, a sea urchin egg develops into a complete sea urchin up to a certain stage of development, despite drastic size reductions. A tail grows back to some lizards, teeth to sharks, hairs to mammals, etc. Worms cut off are reformed. Plants are cut. Tissues are repaired. Cells take on specific shapes. ...

The question of the permanence of living organisms approximates that of individual phenotypic variability. In the same individual, stochastic variability is often measured by right/left asymmetries. We thus assume that the more stable a phenotype, the more symmetric it is. The right/left variations observed for a symmetrical organ are therefore possibly a way of understanding the intensity of canalization in a given organism. In long-lived plants such as large trees, neo-mutations are likely to be very frequent and a genetic mosaic of leaves, branches, flowers and roots is probably to be observed. Despite this, the plant seems to retain a great unity in the form, color and functional organization of its organs suggesting some canalization in their maintenance. On the other hand, it is more difficult to consider variations occurring in an organism over time, since these variations may reflect internal and/or external changes related to the environment or to changes in the biological network's lack of robustness. Variations of several parameters that are assumed to be more fixed (such as temperature in a mammal, for example) may nevertheless represent an approach to the efficiency of the phenotypic canalization.

In unicellular organisms such as yeast, things seem easier and phenotypic variability is often estimated by measuring interindividual variability in a stable strain grown in a homogeneous medium. Such a measure estimates both the permanence of the characters in an organism and in its descendants. With this approach, there are several hundred[104] genes capable of stabilizing the yeast phenotype. They are involved in the organization and integrity of DNA, RNA elongation, protein changes, stress response. They do not appear to be associated with increased mutagenesis. The gene products are highly connected in the protein/protein interaction networks. We find there properties known for the supposed genes involved in canalization which suggests that the question of canalization can actually join the question of the permanence of the phenotype.

Finally, the phenomenon of senescence is usual in living organisms. It is triggered at very variable ages depending on the species and may appear as a constant of the species. This senescence consists in an overall loss of efficiency of the biological network which becomes less efficient. It is then possible to envisage that senescence corresponds to a loss of efficiency of the phenotypic canalization (a sort of "decanalization") with age, since it no longer succeeds in ensuring the permanence of the functional state of the modules due to accumulation of genetic defects or deleterious environmental factors.

An argument in favor of this hypothesis is the observation of monozygotic twins. These twins are very similar and we evoked the fact that this resemblance can be linked to a shared canalization. It turns out that the resemblance between twins decreases with time. This phenomenon is usually considered to be the effect of environmental factors accumulated over the course of life. However, a divergence by progressive loss of the effect of canalization on the phenotype may also explain the lesser resemblance between twins over time.

Modularity and Sexuality

Modularity is an abstract concept that accounts for the existence of integrated but relatively autonomous parts of a living organism modeled as a network.[105] A module can be defined for questions of topography, function, stages of development, etc. It indicates a series of biological units

[104] Levy, S.F. et al. 2008. PLOS Biology, 6: e264.
[105] Wagner, G.T. et al. 2007. Nature Genet. Rev., 8: 921–31.

(molecules, cells, organisms) strongly connected to each other and forming a unified whole, capable of responding coherently to an input. The origin of the organization in modules of the living remains a question not completely solved. It appears to be present in all living organisms and simulation studies suggest that it appears spontaneously with the complexity of the networks.[106]

However, there is a great variability in the organization of the modules. Thus, the proteins of the metabolic networks participating in a given module differ from one species to another.[107] In other words, there are multiple ways of forming modules with the same elementary bricks (here enzymes). If the phenotypic canalization reflects a resemblance between related entities in the organization of the biological network, it implies that there exists a certain heritability of the modules and therefore that the difference between modules increases with the phylogenetic distance. This seems to be the case and it is possible to establish a convincing phylogenetic tree of the living based on the modules of the network of protein-protein interactions.[108] This observation is in agreement with the hypothesis of phenotypic canalization which transmits the organization of the modules.

Most mutations have a deleterious effect. They cause a loss of "fitness" of the body. We are talking about the burden of mutations that accumulate in the gene pool of a given species. The modularity of biological networks seems to reduce this burden. Indeed, one-time changes in a sub-unit are less likely to have systemic repercussions, limiting the burden of mutations. Moreover, for a given module, the strong interconnection of the biological elements makes it possible to compensate for a possible malfunction of a single element. The transmission of the modular structure of the living organism (whatever its cause) is therefore probably essential for the maintenance of an evolutionary line, especially for very complex organisms. The canalization can then be considered as a parameter associated with the robustness of life.

But it is also probably necessary for inheritance itself to be of a modular type, that is to say that it is made module by module and that it does not merely transmit a structure composed of modules. This point was discussed by Maynard Smith and Szathmary.[109] A holistic heredity (which is implied in the paradigm of the genetic program) that would transmit the organization of the network (even modular) as a whole would have a risk of significant systemic effect in the event of a mutation. There would be a risk of network destructuring in a vital way with the extinction of many lines. This effect would be more and more marked with the complexity of the organisms. On

[106] Hintze, A. and Adami, C. 2008. PLoS Comput. Biol., 4: e23.
[107] Peregrin-Alvarez, J.M. et al. 2009. Genome Biology, 10: R63.
[108] Erten, S. et al. 2009. BMC Bioinformatics, 10: 333.
[109] Maynard Smith, J. and Szathmary, E. The origins of life, p. 9.

the contrary, module by module heredity would reduce the risk of systemic contamination and improve the survival probability of the line.

Let us take an analogy and assume that the chapter you are reading is rewritten by moving the sentences randomly from one paragraph to another. The text would remain modular but would certainly lose much of its meaning. On the contrary, if we rewrite this chapter by interposing the sentences within the limits of the paragraphs or by interchanging the paragraphs together, the text would probably remain largely understandable.

According to the theory of phenotypic canalization developed here, sexuality helps to establish a heredity of each module of the biological network. It thus probably reinforces the modularity of the systems over time thus increasing the robustness of the organism vis-a-vis genetic, stochastic or environmental factors. It is then understood that sexuality can give an advantage to the lines in which it appears. It reduces their risk of extinction because it enhances modularity. Phenotypic canalization could therefore contribute to explaining the selective advantage of sexuality at the level of the lineage, without resorting to selection at the level of the whole species, as is generally the case in the literature.[110]

In general, all examples of inheritance of unlimited type are modular. This is the case of DNA (where the modules are genes), language (where the modules are words), culture (where the modules are learned behaviors), and so on. For a heredity to be compatible with a very strong evolutionary potential, its modular character seems to be a considerable asset. Phenotypic canalization is one way of transmitting the phenotype in a modular way, especially in sexual organisms.

Historicity of the Living

Phenotypic canalization actually reinforces the historicity of living beings. It implies that it is not possible to define the future of a genealogy from the present data. It is necessary to take into account all its past because the past is not summed up in its present as suggested by the classical genetic model (which is actually Newtonian). Because of atavism, with phenotypic canalization, it is not possible to predict the phenotypic evolution of a population by the only allelic frequencies present in the populations alone or even by the current frequencies of the phenotypes. It is also necessary to know the phenotypic past of the populations to extrapolate the future. This opinion seems to be in line with current modelling of evolution.[111] However, an explicit model with phenotypic canalization remains to be evaluated.

---

[110] Maynard-Smith, J. and Szathmary, E. The origins of life, Chapter 7.

[111] Gillois, M. Les modèles dynamiques de l'évolution. *In*: Tort P. Pour Dawin.

Genetic Program and Self-Realization

Our increasingly accurate knowledge of the genome has accustomed us to reasoning in terms of transmission of allele frequencies from one generation to another rather than in terms of transmission of phenotypic frequencies. The question of the phenotype is often relegated to the subordinate question, which is that of the genotype/phenotype relationship. Heredity is not, however, defined by the transmission of a genotype but rather by the transmission of a phenotype (Mendel's laws were defined before the concept of gene). And, as we have seen, the genotype/phenotype relationship cannot be considered subordinate. It is therefore necessary to refocus the issue of heredity on the phenotype, which is the case with the concept of phenotypic canalization.

We have seen that, for complex traits, there is an excess of resemblance between relatives compared to what genetics predicts. It is therefore necessary to look for other sources of resemblance. The idea of a non-genetic heredity is therefore not gratuitous and without foundation. But if we abandon the "all genetic" paradigm, we need another causality that cannot be borne by the environment alone. Examples of monozygotic twins and cloning, for example, indicate that there is a strong hereditary cause. The constraining aspect of phenotypic canalization can be considered as a constraint of development and solve this problem.

It is also logically necessary to look for hereditary factors other than genetic when one assumes that genetic information is probabilistic (in essence or by the complexity of biological mechanisms). Genetic information provides information on the manufacture of elemental living materials (RNA and proteins). But it takes a way of arranging materials to build and maintain a phenotype. To answer the question of the inheritance of "organizational" information, geneticists have sought genes for self-organization, whether they be genes for development or behavior. This approach has been very fruitful with the discovery of chaperone proteins, homeotic genes and certain behavioral genes, for example. But if we consider the genotype/phenotype relationship as probabilistic, these genes can only be seen as facilitating agents. They must be the tools to reveal non-genetic information, but they cannot be the ultimate cause. Indeed, one would then return to the notion of a rigid genetic program. However, the rigidity of a program is not compatible with the living (or even with a network model in the broad sense). To this must be added a capacity for self-organization intrinsic to the model itself. And if the network is stable over generations, then there must be information transmitted on the structure of the network itself.

As a result, and as expected, the model of self-organization of life also implies a transmission of organizational information.[112] Let us take the example of Boolean cellular automata. The published works show that it is possible to obtain stable and reproducible structures from simple rules of self-organization. But for this self-organization to be permanent in a given organism and its descendants, these rules must be stable and transmitted from generation to generation. Without these rules, there is no stability of the phenotype in an individual and the offspring. The hereditary transmission of a self-organization program is thus implied in the model of self-organization in biology. The model of self-organization thus implies in essence a phenotypic canalization. One could argue that in the self-organized model, organizational information may be driven by the quality of the cells of the automata, in other words that the rules of self-organization are due to their nature. But then, this would mean that the genes that code for elementary cells are the holders of organizational information. Then we would return to the paradigm of the genetic program.

In all cases where the concept of a strict genetic program is abandoned in favor of a significant part of self-realization, it seems necessary to imagine hereditary information concerning the mode of self-organization of the biological network defining the living organism. We must therefore admit a non-genetic heredity if we abandon the paradigm of the genetic program. Phenotypic canalization is hereditary, non-genetic information necessary to complement both the theory of the genetic program and that of the self-organization of life. It also serves to reconcile the two antagonistic points of view.

[112] This is also the case for so-called stochastic models: Kupiec, J.J. in Kupiec, J.J., Gandrillon, O., Morange, M. and Silberstein, M. Le hasard au cœur de la cellule.

14

Canalization and Instinct

The variability of the instinctual act and the laws governing this variability must be sufficiently well understood to avoid attributing to experience and to learning phenomena which, in reality, are caused by quite different factors.

Konrad Lorenz

Phenotypic canalization is information about the organization of the biological network. But the biological network is not fixed. It changes at any time. It is malleable. In some cases, network change is stereotyped, reproducible and hereditary. This is called instinctive behavior. Instinct is a particular phenotypic character because it applies to the modalities of reconfiguration of the biological network rather than to its basal state. It expresses the dynamics of the network under the effect of a stimulus. Nevertheless, it is a trait that characterizes the organism.

The inheritance of phenotypes such as the instincts or extended phenotypes described by Richard Dawkins is ill-defined by current concepts of genetic programs or self-organization. Instincts are hereditary behaviors resulting in stereotyped, often complex, actions in response to an external or internal stimulus. Instincts are not reflexes.[113] These are acts. They can be exercised empty (in the absence of stimulus) and may have varying intensities ranging from behavioral roughing to repeated action. Their intensity depends on the internal state of the organism and the reaction threshold is variable. The instincts are complex acts, modifying in a coordinated way the whole organism that moves "en masse". They

[113] For a thorough discussion of this question see Lorenz, K. The foundations of ethology, p. 138 and on.

are determined and respond to precise information. Extended phenotypes are recognizable constructs, realized by living organisms in response to instinctive behavior, such as nests, beaver dams, hives, terriers, etc. These are sophisticated accomplishments that also imply a form of operational information for them to be realized.

There is no doubt that instincts and extended phenotypes are hereditary. Ants rediscover each generation how to make an anthill while only the queen has experienced an anthill in her childhood as a princess. The same applies to parasitic worms, spiders, butterflies, molluscs, reptiles, batrachians, fishes and all solitary animals at the time of their birth and which reproduce instinctive behaviors. Take the example of the web of the spider or of the bird's nest. Neither the spider nor the bird learn from their parents the weaving of the canvas or the construction of the nest. This knowledge is innate. It is hereditary and corresponds to precise information as evidenced by the possibility of recognizing an animal to its web or nest. A phylogenetic classification of animals can be established on their behavior. Yet it is difficult to imagine what genetic program can produce a canvas or a nest. It is even more difficult to admit that a DNA sequence codes for the cuckoo's migratory path! Certainly, "behavior genes" have been described as, for example, the FosB gene and maternal behavior in mice. But if such genes are needed (just as a beak is needed to build a nest or a wire to weave a spider's web), they do not explain instinctive behavior on their own. They are only its material support, its proximal cause. Similarly, the self-organization that would lead each generation to reproduce such an elaborate and stable canvas or nest is difficult to admit. Here too, environmental factors (the quality of the branches in a nest or the resistance of the thread of the cobweb) certainly act on the formation of instincts or extended phenotypes, but these elements are also not sufficient in themselves to be explanatory. They are only contingent.

In total, it seems unlikely that a more complete knowledge of genetic or environmental information definitively solves the question of instincts without resorting to other heritable factors. On the contrary, phenotypic canalization, defined as the sharing of a multi-scale procedure between related parties, is perfectly in line with the transmission of complex phenotypes such as extended phenotypes or individual or social behaviors. Canalization as information on the operating mode can probably carry subtle phenotypes, affecting the organization of the living and its behavior.

Instinct requires an organism capable of apprehending its external environment and reacting by reorganizing its biological network. One might imagine that it develops particularly in higher animals that have a significant motor activity rather than in plants or monocellular species. However, one should not confuse the need for a central nervous system and instinctive behavior. The support of instinct is not to be sought in the nervous or sensory-motor system. It is hereditary information and the nervous system is only a sophisticated tool by which it reveals itself.

Thus, instinct can be highly developed in animals with a modest central nervous system. What an ant does with a very small brain is truly incredible.[114] The brain of the ants has a volume of 0.5 mm^3 and comprises less than one million neurons. Ants can walk, fly, recognize a lot of information (olfactory, visual, tactile, etc.), build, memorize, calculate a surface, communicate, calculate a route, advancing in a military fashion, growing mushrooms, raising aphids or caterpillars, gardening, etc. They are hygienic. They have strategies for the use of resources and can participate in symbiosis. ... The same is true of bees and termites.

Nor should one imagine that instinct is expressed by a single motor reaction. This is a much broader definition of stereotyped reorganization of the biological network to an internal or external stimulus. This can be the secretion of proteins by a cell, the reorganization of the membrane antigens of a parasite, the secretion of volatile products by plants attacked or the modification of the shape of daphne in the presence of a predator. We can extend the definition of instinct to all the scales of the living and consider the folding of a peptide chain or the spatial organization of a cell as instincts at these scales of the living. In this extended sense of the term, instinct is synonymous with hereditary transmission of the organization of the phenotype. Instinct is then synonymous with channeled information. We will use these two terms interchangeably in the rest of this work.

Instinct imposes itself on organisms as a determinism that could be called "internal" or "constitutive". They have no choice but to respond. From the point of view of the organism, the instinct is very binding. It limits his freedom of self-realization. Thus, most instincts cannot be controlled. They are therefore extremely organizing for the living. They exert continuous, constant and lasting pressure on the organization of the living biological network. In this they have the same properties as the phenotypic canalization.

Instinct is usually directed towards a goal that often defines it. It is teleonomic. It responds to the need for a function for the individual: food, reproduction, escape, social organization, protection of youngs, etc. At the level of the individual, it can be seen as a solution adapted to respond to a given environment. In other words, it is an efficient solution of the organism's biological network response for given G and E values. These functional and adaptive properties of instinct are difficult to understand as such.[115] One might imagine that instinct is most often purposeless and useless, that it is a burden for a given species. But if we consider that the

[114] Passera, L. La véritable histoire des fourmis.
[115] This point is addressed by Lorenz, K. The foundations of ethology.

instinct is carried by the phenotypic canalization, these properties become more plausible. They may be likened to Mayr's[116] somatic program. Indeed, Mayr, while strongly advocating the concept of the genetic program, finally understood that non-genetic information was also needed to construct life.

[116] Mayr, E. Afer Darwin. French edition, p. 52 and 55.

Part 3
Evolution

Natural selection is the mechanism of evolution. This fact, which has been debated for decades, is now well established both theoretically and experimentally.

In the modern version of the theory, the motor of evolution is the mutation that induces a new phenotype in a given individual. It evolves in an environment where it is more or less successful in surviving and reproducing itself. If successful, the mutation is transmitted to the offspring and the phenotype is preserved. If the phenotype also gives a differential reproductive advantage over other individuals living in the same environment, then the mutation and the phenotype will become more and more frequent in the population. As a result, the beneficial phenotype settles in the population with the new mutation. This is how species evolve and adapt better and better to their environment.

Let us see how our way of redefining the genotype/phenotype relationship and the modalities of phenotype inheritance can influence how biological evolution is considered.

15

Limits to Natural Selection

I am of the opinion that a good observer is also a good theorist.

Charles Darwin

The mechanism of natural selection, in its modern version, implies several strong hypotheses.[117] We will mention only two. The first is the existence of a strong and even, if possible, univocal genotype/phenotype relationship. Any loose genotype/phenotype relationship results in a major loss of efficacy in natural selection. The second is that natural selection is carried by the environment. In the event of loss of impact of the environment on the survival of the organism or its progeny, the selection becomes less operative.

Natural selection had to be applied as soon as there were inheritable variations. These variations were certainly numerous at the protocell stage. They were divided according to their size, leaving each daughter cell a random part of themselves. Natural selection at that time was therefore already effective. This effect was certainly very strong in terms of negative selection (death due to maladjustment of the biological system in its environment) but very slight in terms of positive selection because all that could live probably had the material means to do so. There was no competition for resources. On the other hand, the strong loss of living matter through negative selection may have contributed to the promotion of effective protocells that have recycled relatively structured waste from nearby protocells.

[117] For an in-depth discussion of natural selection, see Gould, S.J. The structure of evolutionary theory.

The appearance of replicators has had the effect of controlling the number and quality of the elements transmitted from an ascendant to its descendant. This has been an essential factor in stabilizing the phenotype from one generation to the next. Stable phenotypic lines, becoming more efficient, have drastically increased the efficiency of natural selection. This has enabled natural selection to be applied in a positive way in the event of inter-individual competition for resources. This competition for resources has become more and more prominent with the increase in the effectiveness of the living being to survive and to proliferate.

In the bacterial world, natural selection has been (and can still be) applied extremely strongly for the following reasons:

- There is a very high potential for growth of bacterial populations. We are in typical Malthusian conditions as assumed by Darwin. As a result, selection has countless living units on which to exert its effect.
- The bacterium has a small genome. There is little non-coding and non-regulating DNA. The majority of genetic changes result in a change in the protein or the regulation of its synthesis. A genetic modification usually involves a biological change in the bacterium. There is a very good genotype/phenotype correlation.
- Bacterial reproduction is asexual. Except recombination, the phenotype of the daughter bacteria most often coincides with that of the parent bacterium. Natural selection can thus be applied to stable phenotypes over several generations.
- The bacterium lives in an open environment and cannot protect itself from most environmental variations.

The bacterial kingdom is therefore typically that in which Darwinian selection is at work. This has led to an impressive efficiency of these organisms. Extremophiles are a striking example.[118] There are bacteria everywhere. They can withstand virtually everything: heat, cold, desiccation, radiation, ionic concentrations, pH, etc. Bacteria can be considered as factories where natural selection has gradually improved productivity and the specialization of production.

The appearance of eukaryotes was mainly marked by the appearance of the internal skeleton of the cell. This allowed the cell to grow, to phagocyte, to associate itself with other cells or surfaces, to make a mitotic apparatus (with the corollary development of mitosis and meiosis with separate chromosomes). To organize an internal environment with mini organs (organelles) with separate functions and positions, and finally to give shape to the cell. In the longer term, the eukaryotic cell also prepared the multicellular organism. Among all the innovations of eukaryotes linked to

[118] For example see Gross, M. La Vie excentrique.

the skeleton, phagocytosis was probably the most crucial by modifying the ecological niche of the cell and giving it unlimited nutritional resources: prokaryotes. Phagocytosis is also at the origin of symbiosis and the appearance of mitochondria, chloroplasts and maybe other organelles.

Over the development of more complex structures, selection will lose its omnipotence. Thus, organisms that are more complex (and not more evolved) than bacteria have gradually detached themselves from the impact of natural selection. Natural selection must here be taken in the sense of positive selection, the selection "intrinsic" or negative still being of course as strong (except in man due to advances in medicine). Positive natural selection, however, never lost its rights. It is simply that its effect is less felt. Just as every organism remains subject to the laws of electromagnetism or gravity as an inanimate object, these physical laws do not have the same importance for it.

The reasons for the partial emancipation of eukaryotes with regard to natural selection are numerous and rapidly listed here. They relate in particular to the attenuation of the genotype/phenotype relationship and to the mitigation of the impact of the environment. Among the modifications of the genotype/phenotype relationship, the following evolutions can be mentioned:

- The duplication of the genome. It is certain that even complex organisms can function with a single copy of the genome. This is the case of plants that have a haploid period, sometimes very lengthy, in their development cycle. This is also the case for male ants or drones. The duplication of the genome could be selected to repair the damaged DNA. Nevertheless, its consequence has been that the body has two copies of the same gene and therefore of the same protein. The impact of a mutation is thus reduced since there is an alternative. The very large number of lethal recessive mutations (including in the human population) attests to this. This is a major mechanism of dissociation of the genotype/phenotype correlation.

  It should be noted that the duplication of the genome is costly because it requires the use of resources. It is therefore in itself a source of loss of selective advantage. It appeared with eukaryotes. At that time, these cannibal cells had a lot of resources (in this case prokaryotes) without any predators. This change in ecological niche may have allowed them to be temporarily less under selective pressure. The eukaryotes then developed new organizations, more expensive but acceptable because of the resources available. These new innovations, starting with the duplication of the genome, have been shown to contribute to the relative emancipation of organisms from natural selection, thus initiating a virtuous circle. This phenomenon will continue over time.

- Duplication of genes. Most genes or segments of eukaryotic genes (exons) have been duplicated. As for diploidy, the immediate consequence of gene duplication was redundancy. Thus, the appearance of a mutation only slightly modified the phenotype since a second gene took over. On the contrary, the appearance of mutations, far from modifying the immediate phenotype, has allowed secondarily the creation of new genes by rearrangement of exons and by evolution for their own account of these duplicated genes which become pseudogens and then new genes.

  This was only possible because long chromosomes were compatible with life (chromosomes not attached to the cell wall, good replication efficiency, low impact on the fitness of a large genome).
- Functional redundancy (or more precisely degeneration). The appearance of introns has made it possible to segment the genes into more or less functional modules (often corresponding to protein domains) resulting in a combinatorial evolution of the genes which has greatly increased the creativity of the living. Genetic inheritance has become modular at the gene scale. It was then possible to respond to a biological function in many ways in a large number of cases. Thus, for example, innate antimicrobial defense belongs to ten TLR receptors and several NLR genes in humans. Once these functions are partially interchangeable, the effect of a mutation on a single gene is only partially visible.
- The organization of biological modules. The integration of the biological modules (or functional characters) leads to many negative or positive feedback loops and to an important robustness to the biological functions. The effect of a given mutation on the phenotype becomes even more tenuous.
- Sexuality enabled the mix of DNA of several individuals to create very different descendants of their two parents. The consequence is the absence of permanence of the object that is the subject of natural selection over generations since the phenotypes are labile. The phenotypic effect of an advantageous mutation is in reality rarely visible. Indeed, most of the genomic contexts where the mutation is found are probably not optimal for a favorable phenotypic effect. And even if, by chance, the optimum genomic context is found for a given mutation, it changes from the next generation diluting the impact of innovation immediately. The relationship between the genotype and the phenotype becomes loose, which we discussed in detail at the beginning of this work.

  It should be noted that sexuality was certainly only possible because the natural selection pressure had been mitigated by diploidy, duplications and the development of stable functional modules.

Indeed, in a strong selection model, the probability that a hybrid being is better than the most selected being is very low. Therefore, for sexuality to take hold, natural selection has an attenuated effect and allows fitness losses by hybridization. The peaks in the adaptive landscape of the species must be broad and blunted, resembling more the Auvergne volcanoes than the Alpine peaks.

- Sexuality causes loss of reproductive efficiency in many species since only one sex carries children. Sex is therefore disadvantageous. Its long-lasting and widespread presence in all phyla species also testifies to the loss of the effect of natural selection.
- Sexual selection is also a very powerful selection tool that can counteract natural selection. What is selected is what pleases the partner without it being associated with a clear selective advantage. This is the case with the tail of the peacock, the horns of deer, the color of the fish, perhaps the intelligence of the boys. ...

We see that the virtuous circle of emancipation continues. ... Over time and for all these reasons, mutations have been found to have a lower impact on the phenotype. Natural selection then lost much of its ability to evolve eukaryotic species with sexuality. Ultimately, most mutations have become neutral or quasi neutral with respect to selection as shown by Kimura.[119] This is accepted and must be understood as the demonstration of the loss of impact of natural selection in eukaryotes.

The organization of life into increasingly complex forms has also reduced the impact of the environment on survival and reproduction.

- Biological organisms have been organized as networks whose properties of resilience and robustness (linked for a large part to the complexity of the network itself) make it possible to maintain an overall stability to the system, known as homeostasis. Thus, the living organism, increasing its complexity has also become more apt to survive in changing environments. However, every environment is by definition changing (day/night cycles, seasons, climatic variations, presence of other organisms in the neighborhood, etc.). The most robust living being to the variations of the environment (i.e., the most malleable or the most adaptable) is thus favored and selected. But it is also less subject to the pressure of the environment which ultimately reduces the effect of natural selection.
- The interior milieu. The organisms work together so that their cells depend as little as possible on external hazards. They form an organized mass that reduces the harmful effects of exposure to the environment. This occurs in bacterial colonies with the formation of biofilms. But it is especially in multicellular organisms that the effect takes on great

[119] Kimura, M. Théorie neutraliste de l'évolution.

importance with the development of an interior milieu. This interior milieu protects the cells from external variations in terms of ionic composition, pH, temperature, oxygenation, ultraviolet exposure, etc. The development of means of protecting oneself from the environment (or more precisely maintaining a controlled environment) is even a criterion often considered important for judging the "evolved" or "superior" character of an organism. A particularly visible example of how to reproduce: oviparity with or without incubation, viviparity with or without breeding, etc. The living organism is increasingly protected from environmental variations and thus protects itself from the effect of natural selection. It can be questioned where this mechanism will stop. Can we imagine even more homeostatic organisms than mammals?
- Instinctive behavior also significantly modifies the effect of the environment. The living being reconstructs its environment by searching for niches (terriers, territories), by mobility to avoid the environmental exposure which disrupts it, by the variation of its state (modifications of the surface molecules of the parasites), etc. It often develops an extended phenotype that allows it to control its environment (barrier of beavers, nests, cobwebs, termites, etc.).
- Finally, the living being learns by culture to dominate its environment. Culture is by no means unique to man. It exists in birds like the magpie which uncaps the bottles of milk; In monkeys that wash their food, use tools to catch ants or break nuts; Among dolphins who have developed collective fishing methods; In felines or canids with hunting methods, etc.

Finally, we must add the argument of the number. Natural selection is all the more effective because it applies to a large number of individuals and over a large number of generations. However, the organisms considered to be the most complex or the most advanced (birds and mammals) are rare, the number of their young is small and the duration of their development until the age of reproduction is long. Natural selection has less material at its disposal. Similarly, in social insects, socialization reduces the impact of natural selection by reducing the pool of possible fertile offspring.

In total, over time, a set of mechanisms has reduced the impact of point mutations on the phenotype and the direct impact of the environment on the survival of organisms. These mechanisms did not eliminate natural selection but mitigated its effect. Complex organisms are always dependent on the law of natural selection, but this is less important.

16

Instinct and Evolution

It is neither the form of the body nor that of its parts which confer on an animal a particular mode of life and habits; On the contrary, habits, behavior and all the pressures of the environment have shaped over time the different parts of the body of each animal.

Jean-Baptiste Lamarck

Despite the decreased response expected to natural selection, the higher animals have nevertheless evolved. Birds were diversified about 100 million years ago, with 10,000 species enumerated and distributed on all continents and in all biotopes. Around 65 million years (Ma) ago, mammals saw their mass evolve very rapidly (over a few hundred thousand years), going from 100 g on average to 10 kg.[120] Around –55 Ma, in the Eocene inferior, one sees all the modern orders emerge: rodents, bats, herbivores, euprimates, carnivores and cetaceans. While 31 families of mammals are recorded at –65 Ma, there are about 100–55 Ma and 122 at –34 Ma. The radiation of mammals is therefore both considerable and rapid.

Taking the example of cetaceans, they appeared about 50 million years ago. Cetaceans (especially whales) are few in number animals with reduced progeny and a long reproductive cycle (sexual maturity at about 7 years). It is difficult to estimate the total number of whales but the assumption of very high numbers is unrealistic even before hunting by humans. This small workforce reduces the material to which the selection relates. Yet the adaptation of the whales to the pelagic environment is such that all their organs have been modified for their form and function, with the possible

[120] Hartenberger, J.L. Une brève histoire des mammifères, Chapter 4.

exception of the heart.[121] This is probably also true of bats, large dinosaurs or hominids. It is therefore legitimate to question evolutionary mechanisms that assist natural selection in complex living organisms.

We know that the determination of the phenotype is the result of the GxE reaction norm and therefore that the environment has an active role in the formation of the phenotype. This point is well established by studies on phenotypic plasticity. Thus, contrary to the classical theory of natural selection, the environment is not neutral but active in the construction of the phenotype. However, this remark does not satisfactorily answer the question posed of what is added to natural selection to explain the persistent evolutionary capacities of complex organisms.

In bacteria, we have seen that natural selection is extremely effective because of a strong genotype/phenotype correlation that can be reproduced from one generation to another. In the case of sexual reproduction, the phenotypic canalization allows the transmission of the phenotype, character by character, from parent to child, which ensures the perenniality of the traits and consequently defines certain characters as specific to a lineage. In both cases, the persistence from one generation to another allows natural selection to exercise its role. Phenotypic canalization could thus be an aid to natural selection in complex organisms.

The transmission of a character through phenotypic canalization is possible, but we have seen above that this is normative. A new character therefore has difficulty settling in a population (it is difficult to fight habits!). It will settle all the better as it will appear in a small consanguineous population, allowing the recurrence by chance of the phenotype shared by several ancestors. Its recurrence will also be facilitated by a selective advantage associated with it since the number of descendants carrying the phenotype will be higher in the population. It will also be readily established in a population if it is spontaneously recurrent, appearing in unrelated subjects (and therefore in carriers of different genomes). In all these classic situations of population genetics, the phenotype will then propagate rapidly.

One may wonder in which situations a phenotype may occur recurrently in a non-consanguineous population. One can mention the case of hot spots of mutations observed for certain genes. But it is especially the case of a massive and new environmental exposure which should cause new recurring phenotypes to appear. It should therefore be expected that sudden climatic changes or specific exposures of organisms will result in more pronounced phenotypic changes. In these cases, it is possible that emergence of the canalization assists in the development of new phenotypes as noted for experiments with inhibition of Hsp90. This may explain why some geological eras have been associated with periods of

[121] Lecointre, G. and Le Guyader, H. Classification phylogénétique du vivant, p. 451. Chaline, J. and Marchand, D. Les merveilles de l'évolution, Chapter 13.

high species creation while others have been less creative. Thus, it seems that "the Cambrian explosion" that occurred 570 to 510 million years ago and saw the abundance of new organisms was associated with significant environmental changes.[122]

It must be stressed that it is the phenotypic character and not the genome that is stabilized by phenotypic canalization and selected by natural selection. The genome is only the wagon and not the locomotive, it only accompanies the character. This has been mentioned above with regard to homeotic genes. It is enough for the genome to be compatible (i.e., that the phenotype retained by the selection is in the field of possibilities of the genotype) so that it is preserved. Natural selection does not therefore retain precise genetic polymorphisms, but only non-incompatible polymorphisms. Evolution is therefore not summarized to changes in allelic frequencies as is often suggested in population genetics. Frequence variations of the alleles are only an indirect consequence of frequence variations of the phenotypes, which are themselves due to natural selection and canalization. It is then easy to explain that most genetic polymorphisms have a neutral or quasi neutral character with respect to natural selection, even if the latter is real and active! Indeed, canalization allows the maintenance of numerous genetic variations provided that they remain compatible with the selected phenotype. What is selected by nature is no longer a specific gene (what we have become accustomed to in genetics) but a phenotype whose determinism is complex and includes the entire genome and the environment. Phenotypic canalization thus makes it possible to reconcile the points of view of two apparently opposing evolutionary theories: natural selection and the neutralist theory of evolution.

The phenotypes initially determined by a new environment can be transmitted as well as phenotypes determined by neo-mutations. Thus, any acquired trait, whatever its cause (genetic, environmental or random) may be transmitted by the mechanism of phenotypic canalization as seen in the previous chapters. The appearance and maintenance of characters initially borne by the environment naturally recall the Lamarckian heredity or the heredity of acquired characters. The latter is usually rejected from the first chapters of a book of genetics as a demonstrated fact. It is explained that generations of mice have had their tails cut off without transmission to the offspring of a shorter tail, which demonstrates the non-transmission of acquired traits. It would have been possible to dispense with these experiments by the simple observation that it is always necessary to cut the tail of the breed dogs in spite of generations of tails cut by the breeders. Ritual circumcision or the habit of shaving beards in men are also very demonstrative. It is the same with the persistence of the vaginal hymen in women.

---

[122] Lambert, D. and Rezsöhazy, R. Comment les pattes viennent aux serpents, p. 145.

We have seen that, on the contrary, the experiments reported on canalization suggest that a character borne by the environment can be transmitted to the offspring. However, according to our hypothesis, the characters that can be transmitted by canalization must participate in the ontogenesis of the phenotype of the organism. Indeed, phenotypic canalization concerns the realization of the phenotype but not its accidental occurrence. The canalization allows transmission of the GxE reaction norm. It seems very unlikely that accidental modifications of an already installed phenotype can be transmitted as in the case of experiments of mice whose tails have been cut. They do not in any way affect the norm of reaction.

There is also the traditional question of the giraffe's neck reported briefly by Lamarck but taken as a model of discussion by many authors. According to the usual interpretation of Lamarck's ideas, the giraffe extends his neck to feed on the leaves of the trees. This makes the neck longer and the "long neck" phenotype is then passed on to the next generation. Thus, the neck of giraffes grows by dint of being lengthened over time, much like an athlete acquires muscle with physical training. In fact, it is not at all evident that the neck of giraffes can lengthen under the effect of training. Even if this were the case, the "long neck" phenotype would be an asset over the years, like the example of mice with cut tails, and would not concern the initial realization of the phenotype. But we have seen that it is not necessary to wait for phenotypic canalization to transmit characters acquired after the initial realization of the phenotype. In other words, the children of "Monsieur Muscle" and "Madame Coquette" are not expected to be naturally beautiful and muscular! The canalization is not Lamarckian in the classic sense. It helps to maintain new and recurring traits, acquired by mutation, by environmental effect or by chance, but only to the sensitive period corresponding to the creation of the organism. For the giraffe, what is transmissible by phenotypic canalization is the norm of realization of the neck at the time of ontogenesis, in other words the value of GxE. I leave the reader to decide whether or not this character has been acquired.

Natural selection favors the best adapted phenotypes which become more and more frequent. It is then seen that natural selection and phenotypic canalization are two entangled mechanisms that spontaneously support and reinforce each other, forming a virtuous circle. Giraffes with the longest neck have a selective advantage (they eat more leaves of trees) with consequent greater progeny. The phenotype "long neck" becomes more frequent in the population. It therefore becomes more canalised, which reinforces its frequency. Long-neck compatible mutations are becoming more and more common in the population due to the effect of both natural selection and canalization. If the selective advantage persists, the neck of the giraffes is extended over the generations. It can be seen that according to this process, natural selection and phenotypic canalization form a couple whose individual effect is difficult to distinguish but whose effectiveness becomes remarkable. Waddington spoke of genetic assimilation to express the combined effect of canalization and selection (natural or artificial) on allelic frequencies.

According to the classic process of natural selection, the length of the neck is indirectly due to the height of the trees: the phenotype gradually adjusts to the environment like a tenon to a mortise. But the behavior of eating leaves is also hereditary and transmitted to the calves from generation to generation, along with their long neck.[123] Such a co-transmission of the physical phenotype and the instinctive behavior is also necessary for there to be a selective advantage. Indeed, giraffes have no selective advantage when they eat grass (on the contrary). So is it the height of trees or the desire to eat leaves that pulls on the neck of giraffes over generations? It must be admitted that it is probably both. We see then that instinct and natural selection are forces of selection which must act in a coherent manner, that is to say, in the same sense. Would not Darwin and Lamarck be right both?

If phenotypic canalization and natural selection act together, then we understand that instinct is adapted to the environment, which solves the important ethological question of the coherence between instinct and adaptation to the natural environment. In practice, it is difficult to separate instinct and natural selection since they influence each other as we have just seen. If instinct and the environment did not go together, then "discordant" organisms would have fewer offspring (this would be the case for giraffes who eat grass because there is no tree left or because their behavior has changed). These organisms should disappear quickly because the phenotype is not maintained by selection or by canalization. In other words, it is necessary that instinct and natural selection be coherent.

However, one can wonder about special situations in which natural selection and instinctive behavior are in opposition. There are situations where instinctive behavior seems particularly costly and where its adaptive nature is questionable. This is the case for fish migrations, such as salmon or eels. To reproduce, the European eel undertakes one of the longest migrations observed in the marine environment. It covers the 5,000 kilometers separating Europe from its spawning site, the Sargasso Sea, off the Bahamas. It is possible that the drift of the continents has resulted in lengthening, over time, the course of the eel that has retained its spawning behavior at the place of birth. In this case we see that instinct may be in contradiction with natural selection. Indeed, any eel that would have

[123] For an illustration of this question, see the example of the beak of the flamingo. Gould, S.J. The flamingo smile, Chapter 1. It will be noted that the term of use and non use used by Lamarck can also refer to the instinctive behavior of the animal. In this sense, the phenotypic canalization can be considered as lamarckian (but not in the sense of a heredity of the characters acquired after the realization of the organism).

changed instinct about its breeding site would probably have had a major selective advantage.

Radiations in many species adapted to very different ecological niches which are observed when new taxa appear, also advocate an effect of instinct distinct from that of natural selection. Take the example of cetaceans mentioned above. The oceans were ecological niches already occupied before the appearance of cetaceans. The first mammals that went towards the water were therefore competitors for aquatic resources with animals much more adapted and efficient than them. They had no obvious selective advantage over mammals who had the instinct to stay on land, niche for which they were adapted. There is a feeling that the "aquatic" instinct of the first cetaceans opposed rather than followed the effect of natural selection.

It should be noted that instinct always affects the actual environment in which the organism lives and therefore the effect of natural selection. The instincts of all types modulate the environment, such as food, housing, migration, care for young, survival behavior, exploration of the environment, etc. All these instincts always directly or indirectly alter food resources, exposure to predators, access to breeders, exposure to physical elements, etc. It is then possible to consider that the instinct of the most evolved organisms has sometimes become the main engine of evolution by modifying the ecological niche of the species. Natural selection then served as a tool to transform the species, at the request of instinctive behavior. Instead of natural selection being the *primum movens* of the evolutionary mechanism, it is the instinct that served as a starter. For example, S.J. Gould tells in several of his chronicles the evolution of the thumb of the panda.[124] The sesamoid of the thumb of the panda is hypertrophied, forming a kind of 6th opposable finger, very useful to the manipulation of the bamboo which is its main food. In this situation, the vegetarian behavior of the Panda seems to be the motor of evolution. Indeed the other ursids are carnivorous or omnivorous. What good reason, other than the behavior has caused the diet panda to change and resulted in the phenotypic adaptation of the animal? A priori, the ecology of Chinese bamboo forests must certainly allow the panda to pursue the same diverse diet as his cousins!

We have seen that natural selection has gradually lost its efficiency in eukaryotes with sex, the organization of a very robust biological network, the development of niches, and so on. In all these cases, it is ultimately a matter of changes in the organization of life. We can then hypothesize that these steps, assimilated to the concept of progress of the living, have been carried by a rise in the phenomena of self-organization (transmitted and stabilized by phenotypic canalization over generations) at the expense of natural selection. The balance between the two evolving forces has gradually changed over time. Phenotypic canalization, while using natural

[124] Gould, S.J. The panda's thumb, Chapter 1.

selection to improve the performance of organisms, gradually free the living from the impact of this exogenous biological law in order to make it a tool for evolution driven by endogenous factors. It has domesticated natural selection.

The archetype of selection is the artificial selection made by the breeders (who introduce Darwin's book on the origin of species).[125] This selection is extremely effective because (i) it is potentially very strong (any organism that does not have the desired trait has zero offspring), (ii) it is directed (one or a few characteristics are specifically sought such as the quantity of milk produced or resistance to a disease for cattle, for example), leaving most other traits free, and (iii) it is constant over generations (breeders always look for the same qualities consciously or unconsciously). These conditions are rarely fulfilled in nature. Most often, natural selection is not strong at this point. Most organisms survive as best they can in their environment and a selective advantage gives only minimal differences in fitness. Nor is natural selection oriented toward a single character. The selective advantage of an organism is the result of the effect of many traits on survival. Certain potentially advantageous characters can be associated with disadvantageous characters either by chance or by proximity of the genes which determine them on the genome or by functional correlation between them. Finally, natural selection may be changing. This is the case in Galapagos with the Darwin finches whose beak size varies from year to year depending on the seed availability itself linked to the climate.[126] On the contrary, phenotypic canalization ensures a strong, constant, permanent selection pressure, targeted on a few traits close to that of the breeders.

There may, however, be situations in nature where natural selection is strong, oriented and sustainable when an environmental factor carries the bulk of the selective effect in an ecological niche. This is the case of extreme conditions of cold (mountains) or light (caves) for example. This is also the case for diseases with high mortality such as malaria or plague in humans or exposure to toxic products such as insecticides in mosquitoes or antibiotics in bacteria. In all these situations, evolution is very rapid and leads to very marked phenotypes. But it is especially the case of co-evolutions between two species such as orchids and their pollinating vector, the fig tree and its wasp or parasites and their hosts. In these situations, there is a headlong rush (called the "red queen hypothesis") between the two partners who constantly adapt to each other. This results in phenotypic extravagancies and an increasing dependence between partners.

In these last examples, it is the existence of a strong ecological link between two partners that explains the rapid evolution of each of them. This relationship is in fact only a particularly strong link in the network of

---

[125] Darwin, C. On the origin of species by natural selection.

[126] Grant, P. and Grant, R. How and why species multiply?

ecological relationships between species occupying the same territory. It is then possible to consider that it is the phenotypic canalization, at the level of the ecological niche, which, by reinforcing a specific functional link of the network, is the main engine of evolution. A strong and lasting functional link is nothing but a canalization at the level of a network of interactions. We can compare the extraordinary sexual dimorphisms observed in some animals.[127] The two sexual phenotypes of the same species are indeed mandatory partners who have developed a strong functional link for reproduction and careof the young.

The instinct of eating leaves forces the giraffe to live in a new ecological niche and redefines its habitat, diet, predators, etc. It is by eating tree leaves, unlike all other gazelles that the giraffe has acquired a long neck. Similarly, living underground, the mole is forced to drastically adapt its organism when there is no obligation other than instinctive for it to do this. It may be said that instinct actually forces the ecological niche occupied by the species. The instinct then causes the species to be selected on new environmental parameters and to diverge over generations. Natural selection thus becomes the mechanism by which instinct shapes the animal. Natural selection becomes a tool used by instinct. This tool is all the more effective as instinct exerts a constant and lasting effect over generations as the breeder exercises an artificial selection. It imposes itself permanently on the animal which is, in a sense, a prisoner. The species can only adapt to its new niche, defined by its own instinctive behavior.

Instinctive behavior domesticates in some way the evolutionary force of natural selection. It is seized on to create new species. This is all the more true as the species are more evolved, with elaborate and marked instinctive behaviors. Natural selection becomes a means rather than an imposed law from the outside. This change is essentially comparable to the revolution of the living being with enzymatic proteins. These catalyze chemical reactions which are thermodynamically possible but rare under normal conditions. They help the reaction to occur and enable the formation of highly efficient and complex metabolic chains. These metabolic chains are the basis of life which needs to stay away from thermodynamic equilibrium. Enzymes do not thwart the laws of chemistry. They exploit them by domesticating them. It may be said that instinct does not counteract natural selection either. It exploits it. And it reinforces it because instinct is constant over generations, it is exercised on defined characteristics and it has a very strong effect on the survival of the animal. We have seen in the introduction that evolved animals are bad materials for natural selection. On the other hand, they have strongly developed instincts. It is then possible that their evolution is kept active by the effect of this instinct which guides natural selection.

---

[127] See the well documented book by Judson, O. Manuel Universel d'éducation sexuelle à l'usage de toutes les espèces.

There are in fact many known examples where instinctive behavior acts as a factor in selecting the phenotype. This is the case of sexual selection, which is no more than a choice behavior of the sexual partner. The behavior of related parties also provides an environment that can select a phenotype. A very nourishing mother will select a "well nourished" phenotype in her children. An aggressive sibling will induce a phenotype of aggressiveness or submission in his siblings. There are many examples of behavior-induced phenotypes in animal communities (monkeys, wolves, insects, etc.). In these examples, the behavior that modulates the phenotype is carried by relatives, but it can also be carried by the individual himself. For example, the leaves of the trees form an environment for the other leaves, hiding the light and causing them to orient themselves differently, resulting in the general shape of the tree. The consumption of specific foods such as eucalyptus in koala, bamboo in the panda or certain fungi in ants influence their metabolism as a whole. Extended phenotypes such as nests or burrows are another example. Reproductive behavior will also affect the actual environment of the organism (e.g., quest or waiting behavior). In fact, any instinctive behavior changes the environment of the organism directly or indirectly.[128] It can be said that to a large extent the organism surrounds itself. It is therefore not shocking to imagine generalizing the idea that instinctive behavior can participate in the selection of a phenotypic character, using the power of natural selection.

[128] For a more detailed discussion of the impact of the organism on its environment, see Lewontin, R.C. The triple helix, Chapter 2.

17

Biological Function

The finality is like a woman without whom the biologist cannot live but with whom he is ashamed to be seen in public.

Ernst Wilheim von Brücke

There is a tendency towards socialization in most taxa: insects, fish (which form shoals or associate in groups), molluscs (corals), birds (the emperor penguin is particularly exemplary!) Mammals (wolves, monkeys, rats of the desert, dolphins), etc.[129] We can even consider that the multicellular organisms are derived from a socialization of ancestral cells. Some unicellular eukaryotes currently express this tendency. It is also found in bacterial colonies. Socialization thus seems universal. It reflects the natural tendency to the combinatorial self-organization of nature that we observe from subatomic particles to human societies and ecosystems. The most evolved animals in terms of socialization are social insects (bees, ants, termites). These animals have highly developed instincts. They communicate, they build shelters, they organize a social life with the sharing of work and the organization of collective tasks, they know how to respond to a disruptive factor, and so on. Social insect communities can largely be viewed as fully-fledged organizations.

The socialization of insects is little like a Darwinian system. The breeders are reduced in number, limiting the material to which the selection applies. Most individuals accept sterilization. Individuals are not competitors for resources, but rather collaborate, share food, associate to perform

[129] See for an introduction: Cézilly, F., Giraldeau, L.A. and Théraulaz, G. Les sociétés animales: lions fourmis et ouistitis.

complex tasks, limiting the impact of a selective advantage-bearing trait over the survival of the best adapted insect. It is argued that the benefit of socialization for each is greater than its absence. The anthill would provide a beneficial environment such that ants would not "benefit" in using their selective advantage in a personal but only in a collective way. But natural selection is typically between individuals sharing the same ecosystem. To say that an individual has a selective advantage in the absolute does not make sense. The selective advantage is relative to the other individuals and a society should on the contrary exert very strong selective pressures. Just see what happens in human societies! We also invoke a calculation of benefit for the kinship (Hamilton's law) where there would be an advantage of transmission of genes (not phenotypes) in certain situations of particular sexuality such as in the ants. However, ants function in society even if the queens are not related. Termites have a classic male-female sexuality but the workers still accept sterilization, which is by definition incompatible with the idea of natural selection. Finally, there are societies of ants that help each other on a continental scale,[130] which is not yet compatible with the idea of natural selection. Socialization thus seems largely non-Darwinian.

However, eusocial insects have evolved towards the definition of castes where individuals are highly specialized. They are adapted to perform specific functions. Breeding individuals are winged to disseminate societies. Queens are gigantic individuals whose almost exclusive role is to lay eggs. The workers are responsible for the breeding and maintenance of the nursery as well as the supplies. The soldiers defend the city and the workers' columns. Each caste corresponds to a typology perfectly adapted morphologically and physiologically to its function. However, it is not natural selection that has adapted these individuals since, for the most part, they do not reproduce.

A selection must therefore be made not only at the individual level but at the anthill level. It is well understood that ants that are highly adapted to a given function provide a selective advantage for the anthill as a whole but not for themselves, unless indirect advantages are taken into account. In many respects it may be considered that it is the anthill which is the object submitted to natural selection. The installation of the queen and the hatching of the first eggs can be seen as the ontogenesis of the anthill. Its development and its disappearance can also be assimilated to its growth and its death. In fact, the anthill meets the criteria of a living organism seen in the first chapter. It will be objected that it is the queen who reproduces herself, and not the ant-hill. But then, according to the same reasoning, a human being is not an organism since it is his gonads that reproduce and

[130] Passera, L. La vraie histoire des fourmis, p. 217.

not him![131] We see that it is necessary to consider an anthill as a scale in its own right, which also shows a hereditary self-organization.

But then how does the anthill exert an evolutionary selection process on individuals? In reality, the adaptation of social insects does not occur vis-à-vis an environment but with regard to a function. This function is not defined on the scale of the individual but on the scale of the animal society. For the ant, its evolutive engine is its instinctive behavior which prompts it to perform a function. And this function only makes sense at the community level. It thus appears that eusocial insects provide an example in which instinct is the principal evolutionary force of the individual. This force is the tool by which the anthill regulates the activity of individuals so that they perform a function in the most efficient way possible.

In general, the effect of natural selection is exercised from a hierarchical level of life to the other, from top to bottom. As we have seen, the organization of the biological network at level i of the hierarchical scale of life consists in the arrangement of modules which are other than the biological units of the i–1 level. The biological units of level i–1 also correspond to functions for level i. This assertion is supported by the fact that topological and functional networks are superimposed; in other words, a biological function corresponds to a module of the biological network of living organisms.[132] The processes of appearance, stabilization and improvement of a function then correspond to the processes of setting up, reinforcing and self-organizing a module of the network of level i. The phenotypic canalization by stabilizing the organization of the network and by accentuating the modularity of the network, thus contributes to the definition of biological functions.

It is understood that by structuring the biological network, phenotypic canalization forces the organism of level i–1 to determine itself around canalised values of the module exactly as the environment forces the organism to adapt. As already mentioned, level i is de facto the ecosystem for level i–1. The constancy of the structure of the network then exerts an effect on the biological units of level i–1, identical in all points to that of a strong, oriented and sustainable selection. And since a module of the network (at any level of the hierarchical scale of the living) can be assimilated to a function for the higher level, phenotypic canalization thus has a selective effect of a functional nature.

At the highest level, phenotypic canalization is defined by instinct in the common meaning of the term. This constrains the mode of operation of the organism as a whole. From what has just been said, we understand that instinct is teleonomic, that is, it is structured to fulfill a biological

[131] At the maximum, it is the Dawkin's selfish gene theory that is implicitly based on a strong Mendelian-type geneotype/phenotype relationship.

[132] Zhao, Y. and Mooney, S.D. 2012. BMC Genomics, 13: 150–60.

function. On the lower scales of the living, instead of speaking instinct of an organ, a cell or a molecule, we usually speak of function because we see these structures "from above". But at each scale, the function (view of level *i*) is nothing but an "instinct" (seen from level *i*–1). This explains why the castes correspond to functions for the anthill but to instinctive behaviors for the ants. The evolution of the castes towards individuals more and more differentiated and functionally efficient translates their adaptation to their environment which is none other than the anthill.

Multicellular organisms are formed from dozens of different cell types. These differentiated cells are highly specialized. The neuron transmits nerve impulses, the enterocyte absorbs nutrients, the plasma cell produces antibodies, the muscle cell contracts, the red blood cell carries oxygen, and so on. Each cell is perfectly adapted to its function. Moreover, just as eusocial insects are born adapted to their function from birth, each cell is specialized even before exercising its role. Thus, an enterocyte is ready to absorb food before even encountering any food bowl. The cells themselves are organized in tissues and organs also adapted to functions. An eye is ready to see before any exposure to light. Neural networks are already preformed before being used for walking in the newborn of a quadruped or smiling in the human newborn. It is difficult to explain all this through natural selection alone.

Let us take the example of the enterocyte, the absorbent cell of the intestine. It expresses the lactase on its brush border. This enzyme allows the digestion of lactose, a carbohydrate in breast milk. It is therefore vital for mammals at birth. However, lactase is of interest only in the enterocyte. No other cell type has the utility. And indeed, only the enterocyte expresses it on its membrane. Thus, according to the orthodox theory of natural selection, the first organism to have acquired by chance a lactase must also acquire it by chance in its enterocytes. Not only a favorable mutation but also a localization of the favorable expression was necessary in the only cell type in which it is useful. This significantly reduces the likelihood that an acquisition useful to the multicellular organism can be retained by natural selection.

Phenotypic canalization reduces this difficulty. The body imposes on the enterocyte to take on a useful function at the level of the organism: the absorption of food. Seen on the side of the enterocyte, this is equivalent to an instinctive behavior (absorber) which imposes itself on it and which is equivalent to an adaptive pressure. If it expresses the lactase it fulfills its function better. A neuron, on the other hand, will no more fulfill its function by expressing lactase. In the first case, the lactase will be preserved as useful to the individual but not in the other case. Thus, phenotypic canalization facilitates adaptation at each level of the hierarchical scale of life, step by step. Like natural selection, it specializes, which corresponds at the cellular scale to a differentiation and at the scale of a society of insects to the formation of a caste.

However, functional adaptation and natural selection must be distinguished. Canalization is an adaptive force but does not involve selection in the sense of "sorting the most fit" because there is no competition between cells within the organism. A given cell line is preserved whatever happens and there is no competition between cells of the same organism. On the contrary, there is cooperation. We are therefore in a system of progress by adaptation but without living loss. It's a bit like a village that has a baker. If he makes good bread, he brings well-being to the village. Otherwise, the village is less happy but the baker keeps his shop because it nevertheless provides a benefit to the villagers. It is better a bad baker than no baker at all! And the baker progresses most often in the making of bread because the village informs him for each batch of the progress made. Similarly, the canalization advances the function of i–1 level elements in an integrated system of level i. It improves efficiency but does not compete between hierarchies i–1.

To create multicellular organisms it was necessary for the cells to specialize and associate and that most give up sexual reproduction as is the case of insect castes today. This renunciation is completely anti-Darwinian. It cannot be due to natural selection. On the other hand, it may be due to phenotypic canalization if one considers that it is not based on competition but rather protects the cell belonging to an organism that contributes to a higher hierarchical level. Moreover, separating functions into modules corresponding to differentiated biological units, explains that certain cells (or certain eusocial organisms) have renounced the function of reproduction in favor of another function. If this is the case, the phenotypic canalization is very powerful since it can counteract natural selection. Finally, phenotypic canalization promotes transmission to the offspring of functionally efficient typologies at each level of the living hierarchy. Thus, unlike natural selection that is only at work at the level of the whole organism, it can carry functional improvements directly at all levels. The division of functional tasks between macromolecules within the cell is also illustrative.

Over time phenotypic canalization reinforces the structure of the network. Thus, for multicellular organisms, the phenotype of a cell type modifies the environment of the neighboring cell type which can then adapt to it. There is therefore rapid coevolution between cell types. Thus, a neuron in contact with a cell that produces a phenotype of transparency can rapidly produce an eye. Canalization makes it possible to preserve and accelerate this co-evolution and therefore behaves as a cohesive and innovative force by co-adaptation for the living. We have already seen an example about the co-evolution of animal species that are ecologically dependent on one another. The strength of the functional link between organisms (at all levels) is the evolutionary force that leads to co-evolution.

In general, the canalization is a cooperative force. It ultimately helps in the constitution and structuring of biological networks. A network is nothing other than a system of functional links between various units that associate to form a coherent whole. Talking about a network is talking about cooperation. Our Western thinking has put much emphasis on competition between organisms by neglecting the necessary cooperation between them. However, a co-operative force is essential to explain the living at all hierarchical levels. The biological network model and the phenotypic canalization that stabilizes it rebalance the vision of the living being by giving an important place to the cooperation.

Phenotypic canalization can maintain links in a network. It thus helps the formation of the $i + 1$ level in the hierarchy of life and thus contributes to the establishment of this hierarchy. It explains that ants of different castes live together without fighting each other and that anthills exist. It explains the rapid evolution of organs using several cell types. It may explain the relatively rapid stabilization of the molecular symbiosis at the origin of life. ...

If we take the function at the i–1 level, it does not make sense. A retinal cell does not "know" what the organism to which it belongs is and does not "know" what use an eye has (as none of us can comprehend what humanity is!). On the other hand, on the scale of the organ "eye", the presence of photosensitive cells makes sense. At this hierarchical level, a "vision" module will be created which maintains the retinal cell in a differentiated state. The cell "experiences" an "instinct" which causes it to collect the luminous information, but for it the luminous information does not have the meaning that the eye gives it. On the other hand, the eye imposes on the retinal cell to treat luminous information, because it finds its account. Similarly, on the scale of the molecule, the notion of storage of information does not make sense for DNA but it does for a cell. Step by step, at each level of the living hierarchy, the network organizes and maintains the modules it deems useful for it.

It is quite difficult to understand the notion of spontaneous optimization of a network. In other words, how does the level-one organization choose what is useful to it? This recalls the idea of the trajectory of least energy. In mechanics, a body follows an optimal trajectory. Thus, an object falls vertically and does not make convolutions in space before landing on the ground (except the pretty dead leaves!). Yet the object does not foresee what is the shortest possible trajectory. He chooses it "spontaneously". This trajectory is in fact optimized at each moment, the object choosing at any time the solution of least action. Thus, from point to point, it describes the shortest trajectory.

In a comparable way, the living must necessarily optimize their self-organized network spontaneously. They explore "possible" solutions and adopt a local optimum which is considered to be the most favorable. For example, if the photo-receptor activity in the pigment cell of the retina is beneficial to the whole system, it will keep the connections that lead to its activation. On the contrary, if the photosensitivity appears on a liver cell, the system will not create a reinforcement of the synthesis activation loops of rhodopsin (pigment used for photosensitivity). It is, in a way, a spontaneous and immediate reinforcement of what proves to be favorable, at the very moment when the solution is tested. It will be noted that there is no real causality in such a networked system since the causality chains form loops. There is only self-reinforcement of the organization of the module and thus of the whole network.

This way of thinking the organism as capable of optimizing itself may seem strange at first sight. It is in fact inseparable from the idea of a network which is none other than a system of spontaneous optimization of functional links between separate elements. This procedure consists of a random test followed by reinforcement if successful. This approach by trial and error and selection of the optimum value tested is in fact very general in life and it is, to think about it, the only possible option of creation without involving any finalism of any kind. This mechanism is seen for the reinforcement of neural networks, metabolic chains or interactions in an ecosystem. In the case of the phenotype, the random test is the spontaneous choice made by the organism during its self-realization from genetic, environmental and canalised information. The memory of choice during life and between generations is the canalization itself.

The canalised functional phenotypes cannot then be optimized without being put in a situation of functional experimentation. Thus, how to stabilize the expression of lactase in the enterocyte before the individual encounters lactose? It must therefore be admitted that the lactase was selected initially at birth. It is then deduced that the canalization only takes place after the phenotype has been self-realized and thus partially optimized. This may seem contradictory to the earlier discussion that canalization does not transmit phenotypic changes occurring after ontogenesis. In reality, this is possible if the optimization is simultaneous with the realization of the phenotype, exactly as an object chooses its trajectory of least energy at the time of its fall. The GxE reaction norm is thus reproduced and simultaneously updated through each organism. We find here the idea that canalization only intervenes at the moment of the realization of the phenotype.

If a new element contributes regularly to optimizing the phenotype (such as the presence of lactase at birth), this phenotype will correspond to a relative optimum and be conserved. We have seen that in time phenotypic canalization leads to a phenotypic assimilation of the realizations of the living. The regularly canalised phenotypes become monomorphic, they express themselves more and more early until they are assimilated into the usual phenotype of the biological unity. The lactase is then expressed by the enterocyte in all the mammals and this even before the birth. The neural networks of walking also take place before birth in quadrupeds, and we could thus multiply the examples.

Moreover, like natural selection, phenotypic canalization promotes, over time, the genes compatible with the canalised phenotype and leads to the evolution of the genome. There is genetic assimilation throughout the generations. This term, however, has a slightly different meaning from Waddington's. Indeed, while Waddington took as sole cause of the evolution of the genome the selection (natural or artificial), it must be admitted here that the phenotypic canalization itself is also a cause of the evolution of the genome.

If there is a self-reinforcement of the network at the time of its formation, then the information carried by the phenotypic canalization is not only an identical copy of the ancestral modules as we have assumed so far. Each organism in the process of creating itself brings a spontaneous part of innovation that updates the canalised information. As with tradition, there is a reinterpretation with each generation of the ancestral phenotype from the possibilities arising from the genome, the environment and random events. The realization itself, that is to say the spontaneous optimization of the organism at the time of its ontogenesis, then also becomes a source of innovation. The GxE reaction norm is updated and the updated norm is transmitted to the next generation. If the biological unit is capable of optimization at each level of the scale, at the time of its realization, the phenotypic canalization will then transmit an increasingly optimized network organization. There will be functional adaptation of the organism, over generations, not only vis-a-vis its external environment but also with respect to its internal structure. The phenotypic canalization can then transmit optimizations of the GxE reaction norm acquired during previous generations.

Of course, the mechanisms discussed above do not involve centralized decision-making processes. A cell of the body does not decide to stabilize a function just as the anthill does not decide to produce reproducers or soldiers. It is the network that stabilizes simply by its feedback loops: if a functional state is efficient (respectively unnecessary), it generates loops of reinforcement (respectively no loop) by spontaneous optimization. The functional modules will evolve towards increasing efficiency by individual canalization, each module being assimilated to a phenotypic character, following a process comparable to that of natural selection. It is the organism, in the sense of self-organized network, that has an effect on itself. We are dealing with a spontaneous self-organization.

The optimal choice of self-organization gives a selective advantage since it ultimately translates into optimization of the organism as a whole. In this case, we see that the motor of evolution is no longer the only mutation but the set of parameters that can modulate the GxE reaction norm: the genome, the environment and the chance of self-realization. Natural selection then sorts the most optimized phenotypes and canalization allows the preservation of this optimization from generation to generation, gradually improving the phenotype. The overall process is thus quite different from the traditional view of natural selection.

It should be noted that this selection process fits perfectly with the hierarchical structure of living organisms. Thus, a protein of any cell of any ant caste can improve the efficiency of an anthill in its environment, even if the anthill does not have its own "gene" (in the modern sense). This process allows the perpetuation of the best self-realized phenotypes, thus optimizing the self-realization of the living organism over generations, at all hierarchical scales of life. It does not, therefore, imply multiplying the molecules storing organizational information at each level of the hierarchical scale of the living organism as one might imagine at first glance. The hierarchy of life can then continue without any obvious limit as the increase in size and complexity of the network itself.

If we consider all the hierarchical levels of living organisms, there is always a functional selection pressure transmitted step by step thanks to the hierarchical structure of the network. One could then think that the lowest hierarchical level is the most prone to the selection pressure and that it is then that which adapts the most. However, we have seen that the lower hierarchical levels of living are also the oldest and therefore the most canalised and the most monomorphic. They are less subject to phenotypic variations even if they are still possible. Adaptation is therefore expected to occur most often at the highest levels of organization during a lifetime. For a bacterium, it will be at the level of protein, for an individual at the level of organs and for a society at the level of individuals. The observation seems consistent with this remark. For example, chimpanzees and humans differ in the anatomy and functional organization of their organs, but their tissue, cellular and molecular organizations are almost identical.

These fronts of evolution of the living are not present in the same place from one epoch to another. It seems to have taken a long time for multicellular organisms to appear. On the basis of progressive gains, more and more complex constructions have been able to establish themselves, leading to the hierarchical organization of living things that we know today. Life advances, like the whole universe, by the association of more and more complex modules. It takes time for all the associations between increasingly complex bricks to be realized, from the atom that assembles the subatomic particles to the animal and human societies that bring together individuals with differentiated functions. The ecosystem is the last known level of the organization of life. It provides each species with an ecological niche. The species, in turn, fulfills a function vis-à-vis this ecosystem. Establishing heredity for the self-organization of complex structures has certainly been a necessary condition for the progress of life.

18

Differentiation and Phenotypic Coherence

What the individual perceives as a freedom is nothing less than a necessity from the collective point of view.

Heinz Pagels

Eusocial insects have a society built around castes, that is, independent organisms with a specific differentiation. The ant is capable of a notable phenotypic diversity for the same genotype. Depending on the type of rearing, the same genome can produce sterile or reproductive adults. Males are only used for the production, transport and insemination of spermatozoa. Queens (or gynes) are used for reproduction. Workers perform other duties. Among the workers' caste, several sub-castes can sometimes be distinguished as in the legionary ants where there are minor workers who feed the larvae and the queen; The media workers who ensure the generalist tasks; The long-legged sub-major workers who carry the prey and the working-class soldiers with a head armed with formidable mandibles that defend the colony.[133] The social insects of the same caste have very similar morphotypes and above all very stereotyped behaviors which often gives the entomologists the impression that the insects are kinds of robot-animals. It must be admitted that the work of ants in columns collecting food or massive nuptial flights suggests that their behavior is more collective than individual.

[133] Passera, L. La véritable histoire des fourmis, p. 34.

This instinctive behavior is certainly due to the phenotypic canalization discussed previously. The ants of the same colony are most often from the same genealogical line and therefore there must be canalization phenomena explaining common behaviors. We have seen that these common behaviors are extremely powerful and may explain the specialization of individuals selected for a biological function. However, there is no reason to assume that social insects have more stereotyped behaviors than other living organisms, which are also subject to phenotypic canalization, such as humans, for example. In other words, why are caste rather than non-specialized individuals self-determining in terms of context or intrinsic qualities, such as the choice of trades in mankind?

In order to understand this, insect societies and other animal communities can be compared. Vertebrates tend to group together and form "objects" endowed with new properties. This is the case of a flight of wild geese that draws a pretty triangle in the sky, a colony of wildebeest or a shoal of fish. This is an emerging property of the group. In these situations of gregarious life, the individual responds to an instinct that is beneficial to him personally. The benefit is based on multiple causes that need not be detailed here.[134] Emergence arises just from the fact that individuals tend to react for themselves in the same way because this same way is generally favorable. On the contrary, in societies of insects, the behavior of the individual is not for individual benefit. Its behavior is less sheep-like than stereotyped because of a conditioning of its instinct by the anthill (or the termite mound or the hive). In fact, the behavior is not the same for all the ants since there is differentiation in castes, clearly indicating a conditioning.

To resume the analogy with quantum physics, the stereotyped behavior of social insects recalls the behavior of the identical photons that form laser rays. The laser effect was discovered by Einstein a century ago. The excited atoms, once left at rest, see their electrons "descend" spontaneously to a lower energy level. In doing so, they emit a photon of a given wavelength which will itself be able to drive the stimulated emission of a photon by another previously excited atom, thus forming a chain reaction. All the photons obtained by stimulated emission have the peculiarity of being strictly identical. They form a coherent beam of light. The fact that the photons all have the same properties then makes it possible to obtain a collective effect which is quite interesting, for example the possibility of focusing the laser beam on a minimal surface.

The stereotyped and coordinated character of the behavior of insects of the same caste is comparable to the coherence between particles of a laser beam. The organisms belonging to the same caste have coherent phenotypes and behaviors, forming a unitary collective whole. Each caste acts as a

[134] For further details, the interested reader can refer to the book by Cézilly, F., Giraldeau, L.A. and Théraulaz, G. Les sociétés animales: lions, fourmis et ouistitis.

single organ, carrying out collective actions of far greater dimensions than the mere association of related individuals who act independently as is the case with most animal communities. This behavior is stable throughout the life of the individual and throughout the life of the anthill. It is therefore a lasting coherence in time and space.

What induces the differentiation in castes in eusocial insects is a modulation of the phenotype by the mode of rearing. We can compare this differentiation into castes and sexual differentiation. Sexual differentiation assigns one sex or the other to individuals of the same species. It is most often of genetic origin but not exclusively. Thus, the sex of crocodiles depends on the incubation temperature of the egg, the sex of the clown fish can change over the course of life according to the structure of the social group (like Nemo's dad!). In terms of proximal causality, this assignment depends on sexual hormones just as the orientation in specific castes depends on rearing conditions such as the presence of royal jelly for example. In both cases, the phenotype is oriented towards one or the other phenotype using appropriate molecular tools.

However, in social insects, this orientation is for all characters and for all individuals who will remain permanently "in phase". It is stronger than sexual differentiation. The individuals of a caste always behave consistently throughout their existence, culminating in mass actions. They are caught in a compulsory cooperative system. They can be said to be identical and interchangeable at the scale of the anthill. The insect loses, as it were, its individuality when the anthill assigns to it a caste. On the contrary, a sexed individual, although oriented to a certain (sexual) function, keeps an identity that is never altered. Packaging is only partial. This reminds us in quantum physics of the distinction between bosons which have a tendency to gregariousness associated with their interchangeability, whereas fermions are never identical and interchangeable according to Pauli's exclusion principle. The difference may appear subtle and partly quantitative but it seems to me sufficient to call upon a mechanism of the same nature but more collective than the phenotypic canalization previously discussed, which could be called phenotypic coherence.

Mitosis is the basis of the division of multicellular organisms. Thus, a man is made up of billions of cells all coming from a single original cell: the fertilized egg. The construction of complex organisms therefore relies on cell division to give specialized cells from stem cells (or meristematic cells in a plant). Moreover, differentiation must be maintained over the generations of cells that constitute the living at each moment. It is imperative that differentiated cells maintain a long-lasting phenotype throughout the life of the individual.

The mechanisms of the heredity of cellular differentiation are beginning to be known. They are essentially linked to stable epigenetic marks during mitotic division. It is these epigenetic marks which indicate which genes are usable in each cell. They then allow the expression of specific proteins that will carry the physiological response of the cell in case of stimulation by an external factor (hormone, growth factor, exogenous factor, etc.). It is then logical to propose that phenotypic coherence intervenes not only in the differentiation of insect societies but also in cell differentiation. This phenomenon of differentiation involves, here as in eusocial insects, a stereotyped response to an external stimulus of all cells with the same epigenome. This allows the creation of tissues or organs for a multicellular organism.

Differentiation corresponds to the way in which the "whole" exerts its influence on its parts, imposing stereotyped behaviors on lower level biological units and thus forming coherent groups and endowed with functional properties on its scale (that of the global organism). Differentiation means that there is a type that summarizes the properties of independent organisms, whatever the hierarchical level of the living being. It intervenes for cells (in an organism), for insect castes (in a society). It is emerging for bacterial colonies. One could still go lower down the scale of the living and look for examples of phenotypic coherence for proteins. There are examples of proteins that must all have the same conformation to perform a function. This is the case for proteins of the lens or mucus or for proteins forming networks such as the extracellular matrix. Prion infection in Creutzfeldt-Jakob disease or accumulation of amyloid protein in Alzheimer's disease could be seen as examples of phenotypic coherence dysfunction at the protein level. In these cases, the abnormal protein (whether acquired or innate) induces an abnormal folding of the normal protein, modifying the collective behavior of the normal protein.

The "periodic" cicadas of America have a very long reproductive cycle, 17 years or 13 years depending on the species. Between these periods, the larvae remain underground and feed on the sap of roots. Some bamboos have even longer cycles blooming every 120 years. But what is surprising is that these animals or plants have synchronous cycles. This is particularly surprising for bamboos that bloom on the same date in different parts of the globe. These examples suggest that phenotypic coherence mechanisms may also exist at the scale of a species.[135]

Differentiation participates in the function. Indeed, the simple coherence of behavior of a group of cells brings about the emergence of new properties from a network already constituted. Thus, a cell with photosensitivity is of little use, whereas a retina made up of thousands of coordinated cells becomes a much more interesting functional module. By using secretory/absorbing cells in a coordinated manner, it is also possible to manufacture

[135] For a detailed discussion of these phenomena see Gould, S.J. Ever since darwin, Chapter 11.

an apparatus for excretion (kidney) or absorption (intestine), even if rudimentary. With cells forming interconnections, an information processing apparatus (brain) can be made, etc. The lack of coherence between cells renders such functions inoperative. Similar examples can, of course, be found at other levels of the hierarchical scale of life.

Interestingly, all cells in the body are differentiated (except stem cells). They all belong to a defined cell type. This even seems an obligation since the de-differentiated cells are at the origin of the cancers. Differentiation is therefore obligatory in the cells of a multicellular organism. The return to the wild life of its cells, leads to a disease that destroys the whole organism. If differentiation is compulsory, then one understands that there is a precise and limited number of cell types. Moving the analogy to the castes of animal societies, one can wonder about the effect of the existence of ants or termites which would be resistant to the differentiation in castes (by mutation of a receiver with the royal jelly for example).

If the loss of phenotypic coherence at the hierarchical level i–1 calls into question the structure of the i-level organism, then it must be admitted that the i-level organism is more than the sum of all i–1 level organisms which constitute it. It is all of its parts plus cohesive forces between them. This reminds us of our current view of the atom where both fermions and their binding particle (bosons) are necessary to form a higher-level structure (the proton or the atom, for example).

Interestingly, phenotypic coherence provides an answer to the difficult question of the coordinated evolution of differentiated organisms for a given hierarchical level of the life. It allows us to consider a mechanism for sharing an optimized differentiated phenotype at each hierarchical level of the living. We have seen, with the example of the lactase of the intestinal brush border, that it was very unlikely that a phenotype would appear not only in the genome of a mammal but also in the right cell which had the use: the enterocyte. This difficulty is even more important in eusocial insects where there is an additional level of integration. Thus, for a change to be beneficial to an anthill, this change must occur by chance not only in the right cell but also in the animal of the good caste. We have seen that this problem can be solved by the notion of canalization which allows stabilization of the optimized phenotype step by step through the hierarchical levels. However, an improvement made by a single cell of an innovative ant is not relevant for the higher hierarchical level. The "discovery" must be exploited by all the cells of the same tissue and/or all the ants of the same caste so that a benefit appears and is therefore retained during self-realization. A coherence mechanism between organisms belonging to the same differentiation group must then be considered in order for a benefit to be observed. For a multicellular organism, it is possible to invoke the filiation between cells of the same type and therefore the transmission through the mitosis of epigenetic marks. But this reasoning is no longer possible for the castes of eusocial insects. If an ant worker "discovers" an optimization of the phenotype useful to his caste, this discovery is lost immediately as the innovative ant is sterile. It is difficult to see how the self-organization of the anthill can be accompanied by optimization and therefore evolutionary progress.

The hypothesis of a phenotypic coherence (i.e., of the durable sharing of phenotypic characters between differentiated organisms), is a possible way to solve this problem. If we take the analogy between biology and quantum physics, the differentiation of organisms at a hierarchical level i makes them interchangeable and identical from the moment they contribute to a higher integration scale. The level $i + 1$ serves as a kind of resonance box as a laser cavity where the photons are phased. It transforms an individual organism of level i into a unit of a collective. Moreover, since the photons trapped in the same laser cavity are both the cause and the result of the stimulated emission, the organisms of level i sharing the same type of differentiation are both the cause and the result of the phenotypic differentiation of the cell type. By this mechanism, phenotypic cohesion could explain the collective progress of complex self-organized structures such as animal organs or societies.

If phenotypic coherence exists, it indicates a link between related organisms not just between generations but at the same generation. It must indeed be admitted that the phenotype of a cell affects that of his sister as much as that of his daughter. Therefore, it is not only the previous realization of the parent phenotype as such that conditions the phenotype of the progeny. The canalization appears to us at first sight as heredity because the related organisms follow one another in a temporal line, but the canalization is not directly linked to a chronology. The linking of the phenotypes translates a match and two related organisms living at the same time can influence each other if they are connected in a network. This has already been mentioned for hydra. The analogy between the phenotypic relation between the related and the entanglement between quantum particles discussed earlier is found even more strongly. The phenotypes (respectively the observables) are related to one another although it is not known what value they will take at the time of the realization of the organism (respectively of the measurement on the particle).

19

Speciation

Nevertheless, it is even more difficult for the average monkey to believe that it descends from man.

Henry Louis Mencken

It is estimated that there were 1 billion extinct species and that there are still 5 to 10 million animal species (not including unicorns and dragons!) and 1 to 2 million plant species.[136] A species is a population of interbreeding and resembling organisms that occupy a particular ecological niche. These three properties are complementary and each participates in the definition of the species.

We know that the genome is transmitted and that it participates in defining the phenotype. The phenotype typically depends on the genome and environment (GxE). We have seen that phenotypic canalization has the consequence of increasing the similarity of phenotypic relationship between relatives. Interfecundity is therefore the basis of resemblance by these two mechanisms. We have also seen that the ecological niche shapes the phenotype (through natural selection) and that the ecological niche is linked to the notion of instinct. A species can then be seen as a particular phenotypic solution of living organisms to the three-variable equation: a genome (G), an environmental niche (E), and an instinctive behavior (I) that acts on the reaction norm (GxE). This solution is usually stable because there are interactions between the three parameters that stabilize the phenotype characteristic of the species from generation to generation (Figure 6). This explains why individuals of the same sexed species resemble each other

[136] Mayr, E. After Darwin. French edition, p. 97.

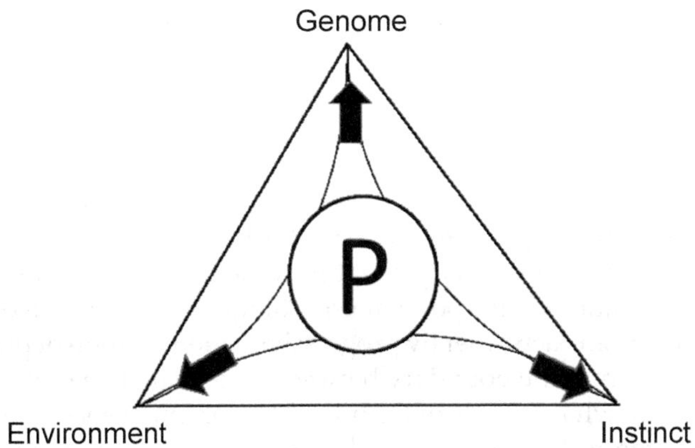

Figure 6. Links between genome, environment and instinct (taken here in the meaning of phenotypic canalization). These three parameters contribute to the development of the phenotype P. Through this phenotypic realization, they echo one another and transmit the updated phenotype over generations. The environment and the genome participate in the phenotype which will then be canalised. The genome and instinct participate in the phenotype that will result in the selection of an ecological niche. Environment and instinct participate in the phenotype that will select the genome by natural selection and assimilation.

when each individual genome is different. This also explains why the three definitions of the species mentioned above are indissociable.

It should also be noted that in a panmictic population with sexuality, all individuals present at the same time have the same distant ancestors. The ancestors having had descendants represent a pool common to this population. The descendants thus share their genes (transmitted by fertilization) and their GxE reaction norms (transmitted by phenotypic canalization). Individuals in the same population therefore share a pool of genes and a pool of phenotypes. According to the transmission methods of the phenotypes by the canalization discussed above, this shared phenotypic pool helps to limit the differences within the species.

Thus, the species corresponds to a population of individuals completely recognizable phenotypically, genetically and ecologically, because they share values close to E, G and GxE and that these three items of information merge into the characteristic phenotype of the species. Although questioned by some, the notion of species is therefore a reality. It is a universal concept for all civilizations and a native living in the heart of the forest very easily agrees on the classification of species with a scientist in his laboratory. A child at an early age recognizes a dog or a cat despite many races. An animal too. By this recognition, functional bonds can be formed between species which can be very narrow, leading to symbiosis, food chains, parasitism ... in short, ecological systems which, without some definition of the species, would have no meaning. The species is therefore a useful and operative concept for all living beings.

Although the notion of species is intuitive and easy to use in everyday life, there are taxonomic difficulties in classifying individuals into separate species or varieties of the same species. The question is then to distinguish the characteristics common to all individuals of the species and those present in only a few. For example, is the shape of the beak of a bird a stable variant defining a species or a variety present in some individuals and subject to frequency fluctuations in the population? This question is expected since the species derive from each other by progressive differentiation of phenotypic characters. Where is the boundary between two related species?

Contrary to what the title of his book "The origin of species by natural selection" suggests, Darwin exploded the concept of discontinuous species by proposing a continuum of phenotypes. In fact, his statement should have led to the disappearance of this concept, at least in its phenotypic definition. The great idea of Darwin's work is that small but repeated changes lead to great achievements if we look back at them in time. This idea is the basis of the theory of natural selection, but Darwin also applied this concept in his studies of earthworms or Pacific atolls. This progressive character is, moreover, essential to the theory of natural selection which cannot otherwise hold true. For Darwin, evolution takes place in small steps and slow changes, from the individual to the variety, then to the species, to the genus, and so on.[137] Nature does not jump. This point of view insists on the absence of an absolute line of demarcation between one species and another. It is fully verified for closely related species that are only distinguished by a few characters such as the color of the plumage of some Amazonian parrots or a different geographical region for the wild mice of Central Asia.

Yet there are many species and individuals are generally easy to classify. Classification is most often based on characteristics considered to be shared by all individuals. These constant characters for a given species are in fact perfectly in agreement with the model proposed here where the phenotype results from information E, G and I (Figure 6). It allows us to rediscover the concept of species that post-Darwinian biology would normally have abandoned.

But the species is also capable of evolution with a readjustment of the parameters G, E and I between them. A mutation with a phenotypic effect can modify behavior and the environment (a mutation coding for a large neck causes antelope to eat tree leaves and live in the savannah). A change in environment changes the genome (through natural selection) and behavior. Thus, a plant transposed into a desert will adapt its genome and morphology (thorns instead of leaves) and its behavior (flowering during short periods of rain). A change in behavior also leads to genetic

[137] Gould, S.J. Structure of evolutionary theory.

and environmental changes (the behavior of cetaceans going to the aquatic environment for example). It is in fact impossible to disassociate the three types of parameters because they are closely related and to say that one outweighs the other is a matter of ideology. Moreover, they form feedback loops (i.e., nonlinear equations) which make the notion of causality between them disappear (Figure 6).

The model proposed here remains compatible with the Darwinian idea of speciation by progressive divergence of the genealogical lines while retaining the concept of species. The realization of living organisms requires an update to each generation of the GxE reaction norm, itself stabilized and transmitted by the phenotypic canalization and thus shared by the group. The canalization thus participates in speciation for separately reproducible lines. Reproductive separation or allopatric speciation may have multiple reasons (geographical, physiological, chromosomal, etc.) which need not be detailed here. In all cases, it leads to a divergence of the genome and/or environmental conditions of the reproductively isolated populations. Since the canalization is done in a straight line in the case of sexual reproduction, the descendants of each isolated population will have a reinforcement of one or a few specific characters which distinguish the founding ancestors, forming varieties then related species up to taxa higher.

In some situations, there is no obvious reproductive barrier (sympatric speciation). This is the case of speciation of fish in closed lakes. In this case, speciation seems to result from different instinctive behaviors that link the divergent lines to different ecological niches. So some fishes are herbivorous and other carnivores. Sympatric speciation also appears to be explained by the model developed so far since we have seen that instinctive behaviors can be evolutionary engines by defining particular ecological niches.

The model of heredity proposed in this work can thus help to reconcile the notions of species easily (by stabilization of characters in a whole lineage) and speciation by divergence (by updating the phenotype in each generation).

20

Phenotypic Innovation

Not only does God play dice but ... sometimes he throws them where they cannot be seen.

Stephen Hawking

If the creation of related species by phenotypic divergence of related lines is easy to admit, the constant progressive and slow character of the divergences has been questioned even among convinced Darwinians like S.J. Gould.[138] It is not a case of questioning the continuity of life, but whether the rate of phenotypic innovations is regular or whether there are accelerations (respectively stagnations) in the formation of new phenotypes. In this case, certain populations of individuals would appear with phenotypic changes no longer small but significant and abrupt on the scale of evolution.

The idea of a momentary acceleration of evolution was mentioned for two main reasons. The first is the observation of an apparent discontinuity in the evolution of species when looking at the palaeontological archives. Important and numerous artefacts probably contribute to this effect, but the discontinuity is so frequent and so marked in some cases that it seems difficult to sweep away this argument. On the other hand, there seem to be periods when numerous radiations suddenly appear (on the scale of evolution) for a given taxon. This is the case of the Cambrian explosion, for example.

[138] Most species show no directional changes throughout the duration of their presence on land. [...] in a given area, a species does not appear gradually as a result of the regular transformation of its ancestors; It emerges suddenly and "completely formed". Gould, S.J. The panda's thumb, p. 209 of the French edition.

Most often, the varieties that we observe correspond to changes that are not very important at the functional level. Changes in color may indeed produce great effects as for birch moth, but they are often of no obvious consequence as for the passerines in our regions. Minor variations in shape often do not lead to changes in function. In short, the varieties or related species are more made for the cabinets of curiosity of the collectors than for the functional innovation.

However, alongside anecdotal phenotypic divergences, there are also innovations such as organs which allow truly novel functions for perception, locomotion, stabilization of the internal environment, reproduction, etc. Let us take the classic example of the formation of the eye. It is a complex organ that requires multiple conditions to bring about a really useful function: the sight. It is clear that the eye did not appear all formed in an individual one fine day.[139] Indeed, we can observe more or less complete and efficient eyes in different species demonstrating if it was necessary that the formation of the eye is a continuous process that takes time. We have seen that the phenotypic canalization helps in the coevolution of several cell types. It can therefore reduce the duration of the formation of the eye. Nevertheless, the formation of the eye poses the following problem: before the eye is efficient, it has no obvious utility and therefore there is no reason to be selected by nature. But if it has not been selected, how can it be efficient? This question refers to the teleological character of the function in biology.

The concept of exaptation attempts to circumvent this difficulty. This word means that a character adapted to a function at a given time can be used (recycled) to another function later. It is thought that the feathers were used for thermoregulation before being useful for flight.[140] The swim bladder of fish is another example of exaptation. This concept of exaptation can be generalized on a molecular scale. The genome of complex organisms contains many genes encoding proteins whose functions can vary from one cell or organism to another. This is the case of an enzyme that has been recycled in the lens because it forms a transparent material.

In fact, any new complex phenotype must rely on what is available and any phenotypic innovation corresponds to a rearrangement of the biological network already existing. There is always necessarily an exaptation underlying an innovation. We discussed the example of Hsp90 mutations revealing cryptic phenotypes. In fact, any mechanism capable of maintaining a stable phenotype by accumulating unrealized "potentials" provides an evolutionary advantage as it prepares future phenotypes. The new phenotype then appears for an ultimate environmental or genetic

[139] For a discussion of the issue see Génermont, J. Les yeux sont-ils des miracles de l'évolution? *In*: Tort P. Pour Darwin; Gasser, F. Les bases embryologiques et génétiques de l'évolution de l'œil: conservatisme et innovation. *Ibid*.

[140] Chaline, J. and Marchand, D. Les merveilles de l'évolution, Chapter 14.

modification, but it has been brewing for a long time (we have addressed this question for Mendelian traits). This exaptation only reflects the very large number of possibilities that the phenotype can take from probabilistic genetic information. It is synonymous with the biological network of phenotypic plasticity. However, exaptation does not resolve the question of the mechanism that moves the biological network "in mass" so that one element of the living (molecule, cell, organ) takes on another biological function and the organism reforms a completely new network around this innovation. It merely expresses the potential for change inherent in each organism.

As suggested by the palaeontological archives, true biological innovations probably appear fairly quickly.[141] If they were actually done in small steps, we should have a very large number of "promising variants" in one species. Given the very large number of steps from, for example, the first photosensitive cell to the eye, the low recurrence rate of a new emergent characteristic in a population and the very large number of possible innovative pathways, each species should comprise a high percentage of such variants. This should be even more evident for species with low number (carnivorous mammals for example) except to say that these species with low numbers are dead ends of evolution, which does not seem to be the case. The emergence of real innovations must necessarily be accompanied by a confusing and lasting bushing of related and/or intermediate forms.

This image of bushing is a good reflection of our current knowledge of the first lines of horses and hominins.[142] It does not, however, correspond to what we currently perceive. If we were in a period of innovation, we should see promising human or equine variants in tens of millions of independent individuals today. This does not seem to be the case (except in the movie X-men!). We observe developmental abnormalities in too many children with genetic or environmental defects but the idea that these children are promising monsters has been abandoned (except perhaps in some cases of high-functioning autism). Their survival is in the vast majority of cases compromised. On the other hand, there are no individuals with an embryo of the eye in the back (which would be very useful!), sensitivity to ultrasound or infra-red waves, organs for retaining microparticles in the air, etc.

The lack of a new, currently beneficial, characteristic of humans (and most domestic species that are sufficiently known and monitored for reliable information) also suggests that innovation is rapid and not very gradually. This is, of course, only a difficult impression to prove and we must be wary of our ability to think about duration at the scale of evolution. It is intellectually difficult to grasp time scales of millions of years. Let's

[141] See the discussion by Gould, S.J. The panda's thumb, Chapters 17 and 18.

[142] See, for example, Gould, S.J. Ever since darwin, Chapter 6; Gould, S.J. Bully for brontosaurus, Chapter 11.

try. The mammal group consists of more than 5,000 genera of about 200 million years old, with animals occupying all ecological niches. According to a rough estimate, it is necessary to imagine the creation of a genus of mammals every 40,000 years on average and thus of species every few thousand years. This figure suggests that we should observe at least some promising phenotypes today among the many known species of mammals living alongside us. But I admit that it is also possible that we do not know how to see them.

Since reproduction is normative, it is unlikely that a promising character is recurrent in a population by chance. This has to be due to a recurring environmental or genetic cause and/or to be selected by nature or by instinctive behavior. Moreover, the observed character must be supplemented by multiple secondary characters which will improve the phenotype in the same direction. Then we come back to the question of a strong selection, oriented and stable over the generations. We have seen that the phenotypic canalization could explain such effects, combined with natural selection, in complex organisms. Instinctive behavior also has a significant effect on the ecological niche of the line. This is the case of the giraffe taken as an example above but it is even more the case of lines that have drastically changed niche like cetaceans, bats, birds, seabirds, etc. In these situations, the behavior associated with niche change has forced these lines to evolve rapidly and in depth.

While phenotypic canalization and natural selection can explain how new lines are established rapidly, there is still a difficulty. It is the engine of it. We know that mutation is the driving force behind the evolution of natural selection. But the mutations are random, rare and ineffective in drastically changing a complex phenotype as a whole. There may of course be exceptions such as homeotic genes whose mutations can profoundly modify the pattern of organism development, even though it is more a "decanalization" model than a small step innovation such as suggested in the mutation/selection model. Speciation by mutation/selection explains quite well the different species such as wolves, coyotes, jackals, dogs. But can we really talk about phenotypic innovations and fundamentally different species (besides, there seems to be some inter-fertility)? It is therefore not certain that the mutation/selection model (with or without canalization) can lead to a drastic modification of the phenotype. In general, biological innovation probably comes only rarely from selection (whatever it may be). Artificial selection, for example, is not very efficient in creating innovations. It only carries the caricature of particular characters. This is perfectly visible for pets or crops. It reveals extreme phenotypes but it is less clear that it is a motor of phenotypic innovations.

Finally, a new organism is not only a phenotypic innovation but several. These multiple innovations must also be coordinated, i.e., coherent and integrated. For example, the creation of a mammal requires the design of both the placenta, the broad pelvis, immune tolerance, the development of a fetal cardiovascular system and so on. Even if we envisage quantities of intermediate steps that are more or less realistic, there is a qualitative jump(s) which must be coordinated and which must be realized in the same promising individual.

It is therefore necessary to propose a mechanism that allows rapid, coordinated innovations that affect the phenotype in depth. This implies a complete reorganization of the biological network of the old organism to create a completely new organism. This corresponds to a change of repository. This mechanism recalls the tunnel effect observed in quantum physics.

Elemental particles are usually kept in a microscopic system and are not free in the universe. This is the case of electrons around a hydrogen nucleus for example. These electrons remain caught in the nucleus by electromagnetic forces. They are said to be in a potential well that they cannot escape because they lack the energy to do so. However, rarely does an electron escape from the potential well. This is the tunnel effect. This phenomenon is possible because the electron can "borrow" for a very short time of the energy and thus jump over the potential barrier and because the probability of presence of the electron (defined by the associated quantum wave to the electron) outside the potential well, although very small, is not zero.

By following our analogical path, we can say that the canalization generates a norm that is similar to a "phenotypic well" that strongly channels the phenotype of the progeny and corresponds to what is called a species. It seems practically impossible that a descendant does not resemble his ascendants, in other words that a dog produces something other than a dog. This is not impossible, however, if one considers that the potentialities of a genome are enormous (as demonstrated by phenotypic plasticity) and that the phenotype is probabilistic. In this case, a dog will occasionally produce something other than a dog. The tipping point (analogous to the energy borrowed by the particle) can come from a mutation, an environmental factor or chance alone. It is possible that it is favored by a transient "decanalization" as seems to be the case with Hsp90 inhibitors. This is not important in itself. As with a loan, the lender is less important than what is done with the borrowed money. It is enough that the tipping point exists in one or a few individuals, whatever the cause, for a change of reference system to appear. There is also no need for environmental mutations or exposures with a strong phenotypic effect for the traditional phenotype of the species to rock. The effect must be quite modest. It should be noted that this mechanism merely translates the plasticity of the living being and does not respond to any orientation whatsoever. It is rare and unpredictable.

This "new thing" has, of course, no chance of being a known organism. This is new, different from everything that existed before. A new, sustainable line is formed, with a complete reconfiguration of the biological network organizing the living. This "block" reconfiguration is possible if one admits that there is an optimization during the self-realization of the organism, which we have seen above. Without this spontaneous optimization, any in-depth alteration of the living structure should be lethal.

However, the new line cannot be stable since the canalization has not yet fixed the phenotypes. Plasticity can then be exercised freely, resulting in the image of a bushing of shapes and innovations immediately after the change of frame of reference. This moment of "re-creation" of a new genus is a moment of bubbling, of strong creativity. It corresponds to the discovery of a new organizational plan. It may therefore happen that several innovations are set up without obligatory links between them. Thus, mammals have developed a placenta, udders and hairs, all elements that are not necessarily related. The birds set up the flight, the feathers, the beak, which were also not necessarily linked. Man developed a large brain, a larynx allowing language and standing. This anarchic creativity is expected in the event of a profound restructuring of the biological network of the organism. It is not expected in case of character-by-character selection over time.

The anarchic and bubbling character of this period of innovation is probably explained by the networking of biological systems. The emergence of an innovation forces the organization to redefine itself around this novelty. For example, standing has forced man to become something other than what he was before. He had to do it and he then discovered a whole series of novelties. It has implemented its phenotypic plasticity to adapt to its new character. It is in fact only a reorganization of the modules in its biological network. In her book on phenotypic plasticity, M.J. West-Eberhard recounts the example of a paralytic monkey in the upper limbs.[143] To compensate for his motor deficit, he stands up to walk in the place of the quadruped position. He invented a new phenotype consequently of a phenotypic modification on another functional module (one speaks of accommodation).

The period of phenotypic innovation seems to open up an exceptional potential for creativity during a certain time window, before the phenotypic canalization recreates one or more wells of phenotypic potential. This may explain the phenomenon of radiation which regularly recurs in the history of life: insects, birds, dinosaurs, mammals, etc. In fact, whenever a new branch appears, it develops into radiations that allow the genus to colonize all possible ecological niches. Everything happens as if the new branch of the

[143] West-Eberhard, M.J. Developmental Plasticity and Evolution, p. 52.

living being had the freedom to follow new instincts. This is despite limited access to resources by other species of other genera already occupying the colonized ecological niche. The period of strong creativity can explain this phenomenon. Conversely, when the window of high variability closes, organisms are enclosed in a new phenotypic well of characteristics that defines the species and channels all individuals in a more determined form.

Very few particles can leave the potential well, at least in the example of the electron participating in a hydrogen atom. Similarly for organisms, very few can come out of their species. A dog almost always gives a dog! Thus, at each major revolution in biology, the emergence must appear unique or rare, which seems to be the case: first proteins (attested by the levorotatory amino acids), first cell, first eukaryote, first multicellular, first bilateral animal, etc. There is an Adam and an Eve for each great innovation and is often given a name (LUCA for the first cell, Bilatera for bilaterally symmetric animals, etc.). This is probably due to the fact that biological networks are extremely robust and that therefore rarely a random change drastically alters the organization of the network as a whole.

We can wonder about the deep meaning of speciation (in the sense of innovation) in biology by pursuing our analogical reasoning with physics. The principle of indeterminacy of Heisenberg is inherent in quantum physics. This uncertainty can be interpreted as follows. The probability for a particle to leave a reference space is very small if one considers a space large enough (one will say the space that is natural for its size) but it is very high if it has a very precise position. In other words, a free particle explores by its nature the space around it, it moves, but at the same time it rarely goes very far. On the contrary, a particle confined in a small space does not have the freedom to move much but when it does, it is to go very far.

In biology, an organism explores the space of possible phenotypes during its self-realization. If this space is not too constrained, then exploration is easy, without great modification of the organization of its biological network, and a great number of varieties are formed which will form related species by reproductive isolation. If, on the contrary, the space of the phenotypes is very constrained (by mutations or an environmental factor or a strong canalization), then the field of possibilities is reduced and the few organisms that have emerged from this phenotypic constraint will have provided a great impulse: they will have redesigned their biological network in depth and they will find themselves with a phenotype far removed from the ancestral phenotype. Too much specialization creates change over time!

Conclusion

We do not know: we are only guessing.

Karl Popper

The relationship between the genotype and the phenotype is, in my opinion, a central issue in modern biology. This work explores this question on the basis of two starting hypotheses. The first is that the relationship between genotype and phenotype is probabilistic. This notion is not original and has been increasingly present in the scientific community for at least 20 years, with the questioning of the analogy of the genetic program *stricto sensu* for the reasons already mentioned. The second hypothesis is that living organisms can be described as a complex network or system. This model is currently widely accepted by biologists. Probabilistic nature of the phenotype and network organization are also easy concepts to associate. In a network, causality is in fact degraded by feedback loops, leading to a looser link between the genotype and the phenotype and from a mathematical point of view to nonlinear equations. The network is also a prototype of a self-organized system. Thus, to a large extent, these two hypotheses are in fact only two ways of saying the same thing.

If the genotype/phenotype relationship is probabilistic, and since observation shows that living organisms resemble each other by progeny, then it is necessary to admit an inheritance of the organisation of the living independent and superimposed to that of the genome. This sharing of information between related parties is thus also assumed, in opposition to the paradigm of the genetic program. We have called this information "phenotypic canalization" to emphasize its pheno-centric (and not geno-centric) character and to indicate its relational nature between relatives. However, this is simply the phenomenon reported by Waddington and others since the 1960s.

The reasoning is quite straightforward and logical and one might wonder if it was necessary to write a book on it. The reason for this is that the model to be achieved has an important explanatory potential if it is developed. The reformulation of the genotype/phenotype relationship makes it possible to update many biological questions. It renews the vision of biological function, instinct, evolution, speciation, differentiation, etc. It is the explanatory character of the model that seems to justify my dwelling on it and which motivated the writing of this essay. The model also makes it possible to unify several biological concepts. It reconciles the paradigms of the genetic program and self-organization; the neutralist and selectionist theories of evolution, or the concepts of stable species and evolutionary divergence. According to the ideas developed here, the elementary character of the phenotype corresponds to a functional module of the biological network. It is also the unit of hereditary transmission channeled. This unity joins the terms instinct and function in a broad sense. Thus, the concepts of gene, phenotypic character, function and instinct are united while taking into account the complexity of the genome and the phenotype. Finally, this work focuses on cooperation between living organisms rather than on their competition.

In this essay we discussed the link between the three types of information that contribute to the realization of living things. Genetics carries structural information (quality and quantity of proteins available). The environment constrains the realization of the living being by the material conditions it provides. Finally, phenotypic canalization provides an operating procedure. In the end, the organism self-realizes and optimizes its phenotype from these three types of information. Sometimes genetics provide structural information that has already been developed. In this case, it is difficult to distinguish between genetics and phenotypic canalization. This is the case for chaperone proteins or homeotic genes. The gene is then an efficient and simple way to organize the living biological network. The hierarchical organization of the living is another way of providing tools already prepared (we should say pre-adapted) for the organization of the living. Indeed, from the bottom up the hierarchy, the living is formatted by the higher level, the lower level biological units providing modules already more or less ready. We have also seen the combined effect of natural selection and canalization on the genome. Canalization and the genome also resonate on the niche occupied by the organism. In sum, all these remarks show that the way of dissecting an organization into three types of information is caricatural. In fact, it is often difficult to separate the types of information that constantly interact with each other (Figure 6). Indeed, this is not very surprising since the realization process of living things is simply to combine all the data into a single object where the three items of information are by definition intermingled. It can be noted that this view rehabilitates the idea of self-realization that produces a phenotype using the three types of information that allow it to be realized. This point of view also gives a central place to the phenotype in biology. It is indeed the object that bears the fate for the next generation. It is through it that the three types of information are updated (Figure 6).

We can argue that this work is only a way of looking at biological questions. However, the ideas developed here, although not demonstrated, are not undemonstrable. They form scientific hypotheses that can be mathematically modeled and experimentally tested, in particular the hypothesis of phenotypic canalization. Of course, the nature of the support of hereditary information suggested here remains to be defined, but this does not prevent us from taking into account the hypothesis and its implications. There are many examples where information support was not known but for which a concept is or has been used. This was the case of natural selection and genetics in their early days. It is still the case today of the instinct in biology, the unconscious in psychology or the quantum entanglement in physics. ...

Until now, the discussion of why the link between genotype and phenotype is probabilistic has been left aside. Let us try to speculate a little on this question quickly sketched in the introduction. It may be summarized as follows: if life is like a lottery, is the circulation by nature random as with a draw from a random list of numbers or is it related to a complex mechanism of which we are unable to follow the movement like a roulette wheel? The probabilistic nature of the link can indeed result from two different types of randomness. The link can be probabilistic in nature, the probability cannot be removed from the biological model as it can not be removed from the model of quantum physics: one can never know more than the field of possible phenotypes (i.e., the random list of numbers). On the contrary, the need to use probabilities to link the genotype to the phenotype may reflect the existence of causal chains that are perfectly deterministic in nature but whose complexity in reality prevents any precise calculation (i.e., the trajectory of the ball of the roulette wheel). In the latter case, it is a chaotic type, in principle reducible, if an extremely precise measure of the state of the system can be obtained at a given moment.

A chaotic phenomenon has two essential properties.[144] The first is the existence of a dependence on the initial conditions that prevents the long-term system from being predicted. This is the famous butterfly effect described by Edward Lorenz.[145] For close initial conditions, a chaotic system will lead to solutions far removed over time. The second property of chaotic systems is the mixture, i.e., the proximity of the final solutions in spite of

[144] This paragraph is largely inspired by Madrid, C. Le papillon et la tornade. In Villani, C., «le monde est mathématique». Editions du Monde, Paris.

[145] "Two states which differ only in tiny quantities can evolve into two totally different states. If there is the slightest error in the observation of a state at the present time, and such errors seem inevitable in any real system, it may well be impossible to make a valid prediction of what will become of this state in the distant future".

distant initial measurements. Indeed, the evolution of a chaotic system is difficult to determine but this evolution will diverge only to some extent because the final results do not cover all the possible values. As a result, the "futures" obtained for two remote initial conditions may prove to be close. These two opposite properties arise from the fact that a deterministic chaotic system associates both stretching and folding exactly like a baker who kneads dough.[146]

Classical physics is chaotic for most of the systems studied. This is due in part to the fact that the studied systems are complex (i.e., with several distinct interactants) and circumscribed in space. This is the case for example of the solar system or balls on a billiard table.[147] While in theory it is possible to calculate precisely the behavior of these systems from simple laws (gravity, friction forces and action-reaction laws), the system is in practice difficult to predict long term. Mathematically, this reflects the fact that the equations formulating the system are not linear and that the solution is very sensitive to the initial conditions of the system. And in reality measurement of initial conditions is always tainted by measurement uncertainty, which necessarily entails important long-term effects.

The uncertainty of the state of a chaotic system thus increases with time elapsed from the conditions considered as initial. While classical physics, which is perfectly deterministic, provides excellent short-term predictability, it does not, in fact, have the capacity to predict in the long run. The classical world is anything but a clock, unless it is reduced to an idealized abstract universe or to consider it on a limited time scale. The result of a roll of dice is in theory perfectly determinable but it is in practice impossible to predict the result!

It is certain that a living organism can be considered as a circumscribed system with multiple heterogeneous elements interacting together. The system is open, subject to multiple environmental variations. It is also subject to internal stochastic variations, which induces randomness in the state of the system at all times. On the whole, even if the evolution of a biological

---

[146] Stretching and folding are found in the mathematical formulations of chaos which express the evolution of a variable from state n to state n + 1 by a perfectly deterministic function. One can take as an example the function xn + 1 = k.xn (1-xn) or the shift application of Bernoulli (where xn + 1 is obtained by multiplying by 10 a number between 0 and 1 and removing the past figure before the comma). In both cases, there is a rapid divergence (stretching) due to multiplication (by k or by 10) and refolding in a small interval of solutions (between 0 and 1, for example, in Bernoulli's application).

[147] "Imagine a particle moving without friction on a line between two walls and undergoing a totally elastic shock. It moves at a constant velocity equal to the initial velocity [..] and we can determine where the particle will be at a given moment if we know its exact velocity. At the slightest inaccuracy in the measurement of velocity, the uncertainty concerning the position of the ball increases with time. If we wait long enough, the imprecision will eventually equal the distance between the two walls. It is thus impossible to predict the position of a particle in the long term". Max Born.

system can be envisaged, in principle, in a completely deterministic way, a biological system is in practice indeterminable. It therefore has all the properties of a chaotic system. The explanatory nature of the genetic program has therefore been denounced because it is impossible to predict the evolution of a biological system in the long term and thus to plan it.

However, we sometimes observe a very large predictability of the phenotype from the genotype as evidenced by the resemblance between true twins. They look almost perfectly alike throughout their lives. To reconcile this predictability observed in the case of twins and the necessarily chaotic model of biological systems, it can be observed that in a chaotic system, for many values of the initial state, the evolution of the system is more or less if we do not take into account the details. This is because, starting from a strong initial divergence, the final solutions are folded within a limited interval. Thus, some solutions, called attractors, summarize a large number of possible futures for a given system. For example, the orbits of the planets in the solar system are approximately the same over the years, although the solar system is complex and unpredictable in the very long term. These recurring attractors or behaviors are frequent and give a certain regularity to the chaos.

In biology, one could imagine giving the attractors an organizing role and solving the question of the resemblance between true twins with them. We have already mentioned such attractors with Boolean automata. Experimental data on cardiac rhythm or encephalographic recording, however, suggest that the presence of attractors is more of a pathological sign whereas a healthy organism is characterized by a chaotic recording without obvious attractor. Above all, the existence of attractors indicates that close end states can be observed for quite distinct initial conditions. If the organism is a chaotic system and if the genome is the primary determinant of the phenotype, then attractors should lead to increased phenotypic resemblance to organisms with distant genomes. An attractor should then increase the resemblance between unrelated organisms but not particularly between related organisms. It should reduce heritability. Using attractors therefore does not resolve the issue of missing heritability.

By contrast, quantum physics is not chaotic but regular. Of course, there is uncertainty about the state of a particle but it is perfectly known and does not depend on the time elapsed from the initial state. It is intrinsic to microscopic reality and is predictable. Thus, quantum physics appears to be much more able to make predictions than classical physics. It is indeed possible to predict at any moment the field of possibilities for a system if its characteristics (nature of particles, experimental system, possible entanglements) are known. Quantum physics, although probabilistic by nature, allows short-term and long-term predictions (which certainly explains the stability of our daily universe!).

Thus, paradoxically, a probabilistic link between the genotype and the phenotype seems more likely to resolve the question of determinism shared by true twins. If the link between genotype and phenotype is analogous to the link between the wave function and the measurement of the state of a particle, then, although probabilistic, the realization of the phenotype is perfectly determined. And if one includes in the probabilities of realization of the phenotype the constraints linked to the matching, then the resemblance between twins becomes explicable.

The notion of phenotypic canalization also appears to be more in line with an essentially probabilistic nature (as in quantum physics) than with a chaotic nature (as in classical physics) of the link between the genotype and the phenotype. Indeed, in a chaotic system, the chain of causality, although impossible to follow in the long term, is perfectly deterministic and leaves no room for any information shared with another independent system. It does not have a non-local effect. The notions of entanglement in quantum physics or phenotypic canalization in biology imply on the contrary a non-local sharing of information between systems. If the phenotypic canalization exists, it therefore argues for a basically probabilistic link between genotype and phenotype. In other words, the probabilistic character of the phenotype and the canalization are two hypotheses, certainly linked by necessity.

In this essay, we have used many notions of physics and more specifically of quantum mechanics. Searching for links between physics and biology is an approach that has been carried out by several physicists, the most famous being Erwin Schrödinger.[148] It seems that Schrödinger's book was an important text for Watson and Crick, directing them towards the search for the molecular structure of DNA. Conversely, for Mayr, "none of the great discoveries in twentieth-century physics has contributed in any way to our understanding of the living world,"[149] suggesting that biology and physics are only distantly related in the phylogenetic tree of sciences. I will not answer that question. Here the approach of linking physics and biology that we followed was not intended to fill the gap between the two disciplines but only to confront the concepts of the two scientific fields by following an analogical reasoning.

Analogical reasoning is only an association of ideas which, by transposing concepts acquired in one domain, makes it possible to generate hypotheses in another domain. It is a powerful tool both for traveling through thought and for conveying complex ideas. As an example and as we have seen, the analogical reasoning between genetics and computing has been particularly fruitful and useful for reflection in biology. Charles Darwin can also be mentioned in "From the Origin of Species", which

[148] Schrödinger, E. What is life?
[149] Mayr, E. After Darwin. French edition, p. 33.

supports his demonstration, explicitly or not, on the analogies between natural selection and artificial selection or between animal populations and Malthusian human populations or between the struggle of the fittest and the invisible hand of Adam Smith.[150] The analogy must, however, be taken for a tool because it cannot prove a thesis. It can only bear it as a diagram illustrates an idea. This is why what is to be remembered is not the analogy itself but the advanced hypotheses and the arguments that support them. The analogical approach developed here is not a demonstration but an illustration.

In this work, the analogy between the subatomic particle and the living one has helped us, starting from the knowledge about the particle, to develop a reflection on the creation of the phenotype. I hope that for the reader, the path that we have travelled by following the analogical reasoning has been rich in new points of view. If this is the case, then one can consider applying the mathematical tools developed for quantum physics to the understanding of the living to prolong this work. This fertility of reasoning was not obvious at the beginning, and it is probable that analogical reasoning between political science and biology would have been less promising (but I have not tried!). One can then question the reasons for this fertility.

Quantum physics describes the microscopic world. It offers a very advanced vision of the mechanisms involved in the realization (in the immediate sense of becoming) of the particles. More precisely, it is a way of approaching the duality between a potentiality and a concrete realization. It seems to me that is why the concepts of quantum physics can yield information for other disciplines of creation.

The question of the realization of an object (in the broad sense) from a potential applies to biology but also to many other fields. For example, we have already mentioned the construction of a house or the production of a recipe for cooking, but we could give many other examples corresponding to all kinds of technological achievements. If we take again the example of the recipe of cooking, the ingredients define the physical elements available for the realization of the recipe. They are analogous to the physical characteristics of a particle (charge, mass, spin). From these ingredients, the possible recipes are most often innumerable. The recipe itself and the know-how learned by the cook correspond to the organizational information. This information is traditional if it is transmitted from generation to generation. It is analogous to the history of the particle and its possible entanglements. Cooking methods and cooking utensils available condition the result of the recipe. They correspond to the physical environment or system in which the particle evolves. We see in this example that quantum analogy only

[150] Gould, S.J. Ever since Darwin, Chapter 4. Gould, S.J. The Panda's thumb, Chapter 5.

allowed us to reveal that creation goes through three kinds of integrated information: structural, environmental and historical.

Quantum analogy must then be applicable to other domains of creation.[151] We can thus imagine an analogy between art and quantum physics. The artist's ideas, emotions or intentions correspond to the structural information. Between the project and the work, there exists the same immense field of possibilities which will be realized during the conception of the work to determine a single object. This is the moment of the anguish of the blank page. Environmental contingency is expressed by the support of the work. For a painting, it will be, for example, the canvas, the brush, the nature of the pigments, etc. The style of the artist can be assimilated to behavior and organizational information. If the style is repeated, it is called artistic school (canalization) which translates an aesthetic community between artists. Finally, some artists will discover another way and launch a new style that will appear as a break or an innovation in the history of art (tunnel effect).

One could also reflect on an analogy between abstract thinking and quantum physics. Ideas form materials. Thought is realization. It depends on the person who manipulates the concepts and is analogical to the apparatus in quantum physics. For example, Darwin (considered here as an experimental device) developed the theory of natural selection (realization) because he was a naturalist because he had travelled around the world because his personal and family history had led him to reflect on the offspring of species and the place of divine creation. It is the person of Darwin who served as an environment for evolutionary thinking to be brought to light on the basis of founding ideas.[152] The ideas that served as the ingredients of the theory were, among other things, the idea that small effects caused great things, the idea that the notion of species has a relative value, the idea of transformism, Malthusian ideas of selection the strongest, etc. The theory of natural selection has been accepted by other naturalists. It was selected and canalised. All evolutionary biologists share it, forming a school of thought (canalization). It is taught at school in a stereotyped way. But some biologists have moved away from the original idea and have proposed new paradigms for evolution (tunnel effect). ...

One could thus develop multiple examples that affect the realization from potentialities. The point common to all is that it is a question of bringing a virtuality to be realized. This implies environmental constraints that require moving from something virtual to a real object endowed with precise properties. Without constraint, the creative act would be unabated. Consequently, the concretization of the object most often involves a creative

[151] For an illustrated discussion of the concept of creation see the beautiful book by Bernard, J. and Donnay, B. Variations on creation.
[152] See Gould, S.J. Ever since Darwin, Chapter 1.

effort that corresponds to ontogeny. This process is a round-trip between the abstract and the concrete, the two being modulated by confrontation. This interaction is part of a know-how that also contributes to the realization. The work is then created by optimization at every moment. It is the arrangement of ideas between them in a logical way or the meeting of the chapters of a book or the musical phrases of a sonata aesthetically. It is necessary to organize a network of intertwined parts forming a harmonious whole. At the end of this creative process, the object has lost all the opportunities of the possible in favor of a single concretization. This process of creation is both extraordinary because it is a question of innovating, of generating the unknown from the already known. But it is also extremely common and altogether banal. In quantum physics, virtual particles are created at every moment. Every moment we see thoughts arise in our minds. And every day, nature draws sheep! In any case, it should be noted that this is always a self-creation. It is in fact the particle, the organism or the thought that creates itself.

We have spoken a lot about information in this text, most often using this term in a common sense, and without really going into the issue. In the theory of thermodynamics, Bolztmann's formula measures the entropy from the N possible microscopic states of a macroscopic system (i.e., containing many molecules) by the formula: $S = k \Sigma_i p(i) \log_e p(i)$ where S is the entropy, k the Boltzmann constant and p(i) the probability of each of the N possible microscopic states. We have seen that the genome expresses the field of all possible that the network can take. It therefore expresses in some sense entropy. In the same way, it is possible to make it correspond to a value Γ corresponding to all the possible N phenotypes from a genome G such that $\Gamma = \Sigma_i p(i) \log p(i)$ where p(i) is the probability of each phenotype i. It would be a kind of genetic entropy for a given genome. The greater the field of possibilities, the higher the value.

In the theory of information, the quantity of information contained in a message composed of N symbols is given by the Shannon formula: $H = -\Sigma_i p(i) \log_2 p(i)$ where *i* is one of the symbols of the message and p(i) its probability of realization. The formulas of Shannon and Bolztmann were brought together and it was proposed by Brillouin that the information be assimilated to a negentropic principle.[153] Entropy is then seen as a lack of information on the microscopic state of a macroscopic system, which naturally returns us to the value Γ. Physical measurement (in the broad sense of classical or quantum physics) is one way of removing this lack of information about the system. In biology, organism formation from the egg can be assimilated to measurement and reduction of indeterminacy by information of the system. The realization of an organism thus amounts to

[153] For more details see Atlan, H. L'organisation biologique et la théorie de l'information, Chapter 9.

informing a system incompletely determined at the outset. This point of view can be generalized to any creative process.

As expected, there is also the idea of information in networks. In a complex network[154] (on the contrary to unorganized systems associating several independent elements without forming a network), the association of the signals between them reduces the entropy of the system (especially if the association is unlikely). The structuring of a complex network is thus associated with the creation of information.

Schrödinger also thought on the notion of entropy for the living. The living, like any material object, tends towards a thermodynamic equilibrium of maximum entropy (in this case, death). Schrödinger remarks that to ensure the permanence of its complex structure, it must compensate for its loss of entropy by the consumption of a negative entropy (and not only of an energy), in other words, of negentropy.[155] If negentropy and information are combined, then we must think that the living person maintains his structure by making and consuming information at every moment. We can suggest here that the chanelled information may play this role. It is this point that makes the essential difference between the living and the inanimate objects and the structural, organizational or environmental data are used only to keep the biological system informed at all times.

An organism is also a machine that is capable of producing information. To explain why this is the case, let's take the example of an enzyme. An enzyme receives the substrates and directs them towards each other, thus allowing catalysis. Catalysis increases the likelihood of a possible but rare reaction. Thereafter, the products of the reaction are released. In addition to its catalytic activity as such, the enzyme directs the chemical reaction by contacting the substrates and disposing of the final products. It can be seen that, apart from any energy aspect of the reaction, an information dimension is involved as soon as there is an asymmetric capacity for fixing/releasing the substrates and products. The enzyme then brings asymmetry to the reaction and displaces the chemical equilibrium. It therefore produces unlikely states and thus increases system information (unlike non-enzymatic catalysis that does not guide the chemical reaction). Another way of saying this is that it keeps the cell away from thermodynamic equilibrium. It acts as a Maxwell demon that sorts the molecules between two compartments and reduces the overall entropy of the system. This action is not related to

---

[154] See Ricard J. L'origine de la complexité biologique. In Maurel, M.C. and Miquel, P.A. Nouveaux débats sur le vivant.

[155] Schröndinger E. What is life? French edition, p. 172. Although less developed and less vulgarized than the idea of aperiodic crystal, it is nevertheless an important notion of the book. It is in fact, according to the author, the idea that justified the writing of his book.

an energy expenditure but only to the functional properties of the enzyme. And these functional properties derive from the fact that an enzyme is already a complex system.[156]

From the scale of the protein to the hierarchy of living organisms, organisms are therefore capable of generating information. Finally, with DNA and phenotypic canalization, they are also able to transmit this information from one organism to another. With time and reliability of these mechanisms, organizations have become increasingly complex, that is to say, more and more informed. They have also become increasingly effective in treating external information with the development of instincts and free will. Ultimately, they bring more and more information to their environment by acting as Maxwell's demons, increasingly effective. To illustrate this last point, it is sufficient to look at the impact of living organisms on the evolution of the Earth (formation of limestone, oxygen, fossil carbon, etc.). Everything happens as if life informs the universe, serving as counterpart to the second principle of thermodynamics.

The concept of heredity of behavior developed in this text refers, as we have seen, to a sort of tradition. It implies an empathy between related organisms. Some metaphysical works have evoked such an idea. I will discuss two to finish.

Rupert Sheldrake proposed that the forms and behaviors of living be defined from morphogenetic fields or morphic fields.[157] These fields include, among others, the form and behavior of organisms. Morphic fields are comparable to universal archetypes that are all the stronger as forms and behaviors are spread throughout the universe. There is therefore a certain resemblance to the idea of canalization. But, contrary to phenotypic canalization, the information carried by the morphic fields is not transmitted solely by progeny. Moreover, the realization of the living is not explicitly probabilistic in Sheldrake's hypothesis. Finally, the morphic fields can carry a heredity of the acquired characters.

The theory developed by Sheldrake, using the notion of field, also brings back to the quantum analogy developed in this work. The field, in its usual sense, is constituted from a physical particle whose presence and its own characteristics (charge, mass) are "signalized" in space. It is actually information about the existing particle that is remotely detectable. A bit like

---

[156] *Ibid.* See also Cunchillos C. Interprétation de la fonction enzymatique des protéines à partir de la théorie des unités de niveau d'intégration. *In*: Tort, P. Pour Darwin.

[157] Sheldrake, R. La mémoire et la matière p. 146 "According to the hypothesis of formative causality, DNA or rather a small part of DNA is responsible for the encoding of RNA and amino acid sequences in proteins, which play an essential role in operation and development of organizations. But the forms of cells, tissues, organs and organisms as a whole are shaped not by DNA but by morphic fields. Similarly, the acquired behavior of animals is organized in morphic fields. Genetic changes can affect both form and behavior, but these patterns of activity are transmitted by morphic resonance. [...] Let us pursue this analogy of developing organisms are connected to similar earlier organisms acting as morphic "transmitters".

a megaphone that announces information on the street. The information emitted by the particle can be detected by another receptor particle. Like a passerby who hears the megaphone. There is a certain resemblance between the transmitter and the receiver. For a particle, it is a question of having a mass or a non-zero charge. For a passer-by, it is about being able to speak and hear the same language. In all cases, an affinity between objects is required. If the information is received, the receiving particle (test particle) reacts according to the given information. The reaction is predicted by physical laws. Similarly, a passer-by will probably react to the megaphone and some parameters will specify the type of reaction. For example, an injunction to perform an action is all the more pressing as the megaphone is close and powerful (in decibels). The effect on the passer of the megaphone can be compared to a force just like the effect of a field on a particle. A force field is thus defined as an informed zone able to set in motion a receptor particle.

The reader will not fail (if not too tired of analogies!) to perceive resemblances to the idea of phenotypic canalization. However, if phenotypic canalization connects two organisms, it is not an exchange of information through communication. It is a special and "personal" relationship between two organisms through their matching. This is an essential feature of phenotypic canalization. Phenotypic canalization is therefore not analogue to a field but rather to an entanglement as we have seen.

Henri Bergson wondered about the intuitive nature of the living.[158] The idea that the living recognizes the living through simple "empathy" would derive from the continuity property of the living being as such (unlike the discontinuity of "living beings"), thus defining a kind of fabric covering what lives.[159] For Bergson, instinct is akin to the modern term of self-organization of life. This position corresponds in this sense to the notion of phenotypic canalization.

In order to prolong Bergson's thought, one may ask where the evolutionary potential of the phenotypic canalization in other words instinct, stops. Using the power of natural selection, he manipulates a

---

[158] Bergson, H. L'évolution Créatrice, p. 53. "Let us indicate at once the principle of our demonstration. We said that life, from its origins, is the continuation of one and the same impulse that has been divided between divergent lines of evolution."

[159] *Ibid*, p. 168: «In either case, whether animal instincts or the vital properties of the cell, the same science and the same ornament are manifested. Things happen as if the cell knew other cells what interests it, the animal of the other animals what it can use, all the rest remaining in the shade. It seems that life, as soon as it has contracted itself into a definite species, loses contact with the rest of itself, except, however, on one or two points which interest the species which has just been born».

Ibid, p. 166: «It is on the very form of life, on the contrary, that instinct is molded. While intelligence treats all things mechanically, instinct proceeds, if we may so speak, organically. If the consciousness that slept within him woke up [...], he would give us the most intimate secrets of life. For it merely continues the work by which life organizes matter, to the extent that we cannot say, as has often been shown, where organization ends and instinct begins.»

very powerful evolutionary force. However, in higher organisms, instincts become numerous and complex. They then form a network of hereditary behaviors. As in any network, the maintenance of stable modules (in this case the most specific and "vital" behaviors such as reproduction or feeding) implies an adaptation of the network.

Bergson opposes instinct and intelligence. He sees two ways of apprehending the world. The fact is that certain lineages of the living have developed free will (mammals, birds) while others have reinforced actions more stereotyped (the social insects). One can then think that insects have solved the question of the management of conflicts between different instincts by the separation of these between different castes. This is particularly true for the separation of reproductive behavior and feeding behavior. As a result, there are individuals with stereotyped instincts, responding to certain biological functions of animal society which is seen as a super-organism whose performance is exceptional. This observation may be generalizable to all insects, including non-social insects. Indeed, in all insects, there is a larval stage and an adult stage (imago). It can be stated in a simple way that during the life of the insect the larval stage corresponds to the function of nutrition while the adult stage corresponds to the reproductive function.[160] The same reasoning can be applied to all multicellular animals with separation of the germ line and multiple somatic lines. The clear separation of operational and informational tasks between macromolecules can also be related to this idea. It can be argued that one of the ways to solve the problems of instinctual conflicts (or functional conflicts, which amounts to the same thing) is to assign different tasks to subgroups of lower level organizations on the hierarchical scale of living organisms, corresponding to differentiated phenotypes. This was the path chosen by the cells, the multicellular animals and especially the insects. As we have seen, each caste or cell or differentiated molecule corresponds to a function or instinct.

In vertebrates, on the contrary, in the great network of various instincts, some of them (defense or exploration instincts, for example) have become looser and more adaptable to resolve conflicts between instincts in the same individual. This malleability of certain instincts has enabled the reinforcement of the most essential instincts such as reproduction and feeding. This explanation is consistent with Konrad Lorenz's observation that domestic animals and humans have less specialization and looser instinctive behaviors, with the exception of feeding and breeding behaviors that appear to be more proeminent.[161]

In doing so, vertebrates have revealed what can be considered as free will, that is an ability to assume behavior in a less stereotypical, more

[160] Gould, S.J. La foire aux dinosaures, Chapter 17.

[161] Lorenz, K. Le comportement animal et humain, p. 135 and following.

flexible manner. The organism was then able to adjust its loose behaviors by learning. Learning by trial has possibly led to consciousness. Learning by mimicry may have been the basis of culture. Culture and consciousness appear (willingly together) in the most evolved animals, living in stable social groups such as monkeys, dolphins or birds. They enable not only vertically but also horizontally to transmit operating modes. They are therefore capable of counteracting the normative power of instincts in higher animals, at least for certain functions.

In Summary

The relationship between the genotype and the phenotype is a difficult question, imperfectly solved today. The study of complex genetic traits shows that a phenotypic character depends on the genome as a whole rather than on one or a few genes. The realization of the phenotype also depends on the whole environment in which the organism develops. However, it is not possible today to predict a precise phenotype from a complex genome and its environment. Until there is evidence to the contrary, the GxE reaction norm that carries out the realization of the organism must therefore be considered as a probabilistic function. This is our starting hypothesis.

A living organism can be modelled as a complex biological network associating various biological units. Among other properties, this network is modular, that is to say that some of its parts can be considered structurally and functionally independent, at least in a first approximation. A module can be assimilated to an elementary phenotypic trait and its functional state to a phenotypic measurement. For each organism there is a large number of elementary phenotypic traits. The combination of several elementary characters forms complex traits like diseases. The set of values taken by all elementary characters defines the individual at a given time.

For each elementary trait, the phenotypic value taken is a probabilistic value that integrates all the genetic and environmental data. This value is updated at all times, allowing the organization to adapt to its external environment and to its internal structural changes. Some of the elementary phenotypic traits, however, appear more stable and others more labile. The most stable values define a higher individual, species or taxon depending on their phylogenetic age.

If the genotype/phenotype relationship is probabilistic, the resemblance between related parties implies that the GxE reaction norm is at least partly inheritable. The transmission of the reaction norm from one generation to the next is referred to herein as "phenotypic canalization". Phenotypic canalization is a heredity of the self-organization of the biological network. The canalization unit is the elementary phenotypic trait which is also a module of the network and which can therefore be considered as a gene in the historical sense of the term. Phenotypic canalization then tends to stabilize the value of the elementary phenotypic trait over the generations, thus limiting the random part of phenotype realization and reducing the impact of genetic and environmental changes on the phenotype in the lineage.

Phenotypic canalization can be seen as a tradition. It is only explained by the related parentage. What is transmitted to the organism in the making is what has already been realized in its lineage. Canalization does not create. It may allow the resurgence of old phenotypic values (atavism) but in most cases, the duct reproduces the parental phenotype. It thus brings a constraining limit to the realization of an organization. As with tradition, each generation can, however, update the GxE reaction norm, allowing for the brutal or progressive innovations brought about by each new living organism at the time of its ontogenesis. These innovations can be of genetic, environmental or simply random origin. The new GxE reaction norm will be canalised if it becomes frequent in the population (by genetic drift, natural selection or recurrence due to environmental factors for example).

The living organism consists of a hierarchy of nested structures: molecule, cell, organism, society, ecosystem ... Phenotypic canalization is exerted at each hierarchical level and must be seen as a hereditary mode of operation at each scale. It corresponds to the mode of folding of proteins, to cellular differentiation, to the castes of eusocial insects, to instinctive behaviors, to extended phenotypes, and so on. In this sense, it is comparable to an instinct in a broad sense of innate behavior applicable to each hierarchical level of life.

The structure of hierarchical rank i sees its phenotypic state largely determined by the structure of rank $i + 1$ which forms its ecological niche. For the level $i + 1$, the level i is on the contrary comparable to a functional module. Phenotypic canalization, by transmitting the information on the organization of the biological network, connects the hierarchical levels of the living with each other. It thus explains the correspondence between instinctive behavior (for level i) and function (for level $i + 1$).

By stabilizing the GxE reaction norm, phenotypic canalization provides stable and strong information over generations that imposes one or a few phenotypes among the immense field of possibilities. The canalization (or instinct) therefore exerts a selective pressure on specific elementary phenotypic characters. It imposes itself on each hierarchical scale of the living and has a functional effect. It can then be seen as an engine of evolution. Its mechanisms are entangled with those of natural selection since instinct makes the environment of the organism at all scales. Its role probably increased with the complexification of the living.

The creation of a living being corresponds ultimately to the information of a non-predetermined biological system by a combination of data from the genome, the environment and phenotypic canalization. In practice, these three types of information merge into one at the time of the realization of the living organism. They also interact with one another over generations, resulting in complex feedback loops. It is then possible to define the species as a group of related organisms sharing their genome, environment and phenotypic heredity. Species can evolve by slow or abrupt modification of one or more of these parameters explaining the mechanisms of speciation.

Phenotypic canalization is similar to canalization described by Waddington. There is thus a whole corpus of experimental data compatible with the hypothesis proposed here. It appears to depend on chaperone proteins, DNA methylation, microRNAs, homeotic genes, and others. However, the support of the information carried by the phenotypic canalization remains to be determined.

Despite this limitation, the two assumptions made here on (i) the probabilistic nature of the genotype/phenotype relationship and (ii) the hereditary nature of the GxE reaction norm, allow for in-depth re-examination of biological issues such as heritability of complex genetic traits, sexuality, the historical nature of life, embryogenesis, the duality between programming and self-realization of life, the permanence of the phenotype throughout life, function in biology, instinct, species, natural selection, phenotypic innovation, differentiation, etc.

Bibliography

Allano, L. and Clamens, A. 2000. L'évolution. Des faits aux mécanismes. Paris. Editions Ellipses.
Atlan, H. 2006. L'organisation biologique et la théorie de l'information. Paris. Editions du Seuil.
Bergson, H. 1941. L'évolution créatrice. Paris. Presses Universitaires de France.
Bernard, J. and Donnay, B. 1999. Variations sur la création. Paris. Editions Le Pommier-Fayard.
Brack, A. and Raulin, F. 1991. L'évolution chimique et les origines de la vie. Paris. Editions Masson.
Cézilly, F., Giraldeau, L.A. and Théraulaz, G. 2006. Les sociétés animales: lions fourmis et ouistitis. Paris. Editions Le Pommier.
Chaline, J. and Marchand, D. 2002. Les merveilles de l'évolution. Dijon. Presses Universitaires de Dijon.
Cobut, G. 2009. Comprendre l'évolution. 150 ans après Darwin. Bruxelles, Editions De Boeck.
Cox, B. and Forshaw, J. 2011. L'univers quantique. Paris. Editions Dunod.
Darwin, C. 1880. De l'origine des espèces au moyen de la sélection naturelle. Paris. C. Reinwald editeur. translated from "on the origin of species by the means of natural selection".
Dawkins, R. 1990. Le gène égoïste. Paris. Editions Odile Jacob. translated from "the selfish gene".
Feynman, R. 1987. Lumière et matière. Paris. Interéditions. translated from "the strange theory of light and matter".
Fox Keller, E. 2003. Le siècle du gène. Paris. Editions Gallimard. translated from "the century of the gene".
Fox Keller, E. 2004. Expliquer la vie. Paris. Editions Gallimard. translated from "making sense of life".
Gargaud, M. 2003. Les traces du vivant. Bordeaux. Presses Universitaires de Bordeaux.
Gould, S.J. 1982. Le pouce du Panda. Paris. Editions Grasset. translated from "the panda's thumb".
Gould, S.J. 1984. Le sourire du flamand rose. Paris. Editions du seuil. translated from "the flamingo's smile".
Gould, S.J. 1993. La foire aux dinosaures. Paris. Editions du Seuil. translated from "bully for brontosaurus".
Gould, S.J. 1997. Darwin et les grandes énigmes de la vie. Paris. Editions du Seuil. translated from "ever since darwin".
Gould, S.J. 2006. La structure de la théorie de l'évolution. Paris. Editions Gallimard. translated from "the structure of evolutionary theory".
Gouyon, P.H., Henry, J.P. and Arnould, J. 1997. Les avatars du gène. Paris. Editions Belin.
Grant, P. and Grant, R. 2008. How and Why Species Multiply? New Jersey (USA). Princeton University Press.
Gribbin, J. 2006. Le chaos, la complexité et l'émergence de la vie. Paris. Editions Flammarion.

Gribbin, J. 2009. Le chat de Schrodinger, physique quantique et réalité. Paris. Editions Flammarion.
Gross, M. 2003. La vie excentrique. Paris. Editions Belin.
Hartenberger, J.L. 2001. Une brève histoire des mammifères. Paris. Editions Belin.
Histoire Naturelle. 2012. Paris. Editions Flammarion.
Judson, O. 2004. Manuel d'éducation sexuelle à l'usage de toutes les espèces. Paris. Editions du Seuil.
Kimura, M. 1990. Théorie neutraliste de l'évolution. Paris. Editions Flammarion.
Klein, E. 2004. Petit voyage dans le monde des quanta. Paris. Editions Flammarion.
Kupiec, J.J. and Sonigo, P. 2000. Ni dieu ni gène. Paris. Editions du seuil.
Kupiec, J.J., Gandrillon, O., Morange, M. and Silberstein, M. 2009. Le hasard au cœur de la cellule. Paris. Editions Sylepse.
Kupiec, J.J. 2013. La vie et alors? Paris. Editions Belin.
Lambert, G. 2003. La légende des gènes. Paris. Editions Dunod.
Laloe, F. 2011. Comprenons-nous vraiment la mécanique quantique? Paris. CNRS Editions.
Lecointre, G. and Le Guyader, H. 2001. Classification phylogénétique du vivant. 2ème édition. Paris. Editions Belin.
Lewontin, R.C. 2003. La triple hélice. Paris. Editions du Seuil. translated from "the triple helix: gene, organisme and environment".
Lorenz, K. 1970. Le comportement animal et humain. Paris. Editions du Seuil. translated from "studies on animal and human behavior".
Lorenz, K. 1984. Les fondements de l'éthologie. Paris. Editions Flammarion. translated from "the foundations of ethology".
Louis-Gavet, G. 2012. La physique quantique. Paris. Editions Eyrolles.
Lovelock, J. 2010. La terre est un être vivant. L'hypothèse Gaia. Paris. Editions Champs. translated from "gaia: a new look at life on earth".
Maurel, M.C. 1997. La naissance de la vie. Paris. Editions Diderot.
Maurel, M.C. and Miquel, P.A. 2003. Nouveaux débats sur le vivant. Paris. Editions Kimé.
Maynard Smith, J. and Szathmary, E. 2000. Les origines de la vie. Paris. Editions Dunod. translated from "the origin of life: from the birth of life to the origin of language".
Mayr, E. 2006. Après Darwin. Paris. Editions Dunod. translated from "what makes biology unique? considerations on the autonomy of a scientific discipline".
Miquel, P.A. 2008. Biologie du 21ème siècle. Evolution des concepts fondateurs. Bruxelles. Editions De Boeck.
Morange, M. 1998. La part des gènes. Paris. Editions Odile Jacob.
Mouchet, A. 2010. L'étrange subtilité quantique. Paris. Editions Dunod.
Omnes, R. 2006. Les indispensables de la mécanique quantique. Paris. Editions Odile Jacob.
Passera, L. 2006. La veritable histoire des fourmis. Paris. Editions Fayard.
Pigliucci, M. and Müller, G.B. 2010. Evolution, the extended synthesis. Cambridge (USA). Massachusetts Institute of Technology Press Book.
Renck, J.L. and Servais, V. 2002. L'éthologie. Paris. Editions du Seuil.
Ridley, M. 1997. Evolution biologique. Bruxelles. De Boeck Editions.
de Saint-Exupery, A. 1943. Le petit Prince. Paris. Editions Folio.
Schilthuisen, M. 2002. Grenouilles, mouches et pissenlits. Paris. Editions Dunod.
Schrödinger, E. 1990. L'esprit et la matière. Paris. Editions du seuil. translated from "mind and matter".
Schrödinger, E. 1993. Qu'est-ce que la vie. Paris. Editions du Seuil. translated from "What is life?".
Selleri, F. 1986. Le grand débat de la théorie quantique. Paris. Editions Flammarion.
Sheldrake, R. 1988. La mémoire de l'univers. Paris. Editions du Rocher. translated from "the presence of the past morphic resonance and the habits of nature".
Teyssèdre, B. 2002. La vie invisible. Paris. Editions l'Harmattan.
Tort, P. 1997. Pour Darwin. Paris. Presses Universitaires de France.

von Uexkull, J. 2010. Milieu animal et milieu humain. Paris. Editions Payot et Rivages.
Wagner, A. 2011. The Origins of Evolutionary Innovations. New-York. Oxford University Press.
West-Eberhard, M.J. 2003. Developmental Plasticity and Evolution. New York. Oxford University Press.

Thanks

Many thanks to Jean-Marc Victor and Alexis Mosca for reviewing the manuscript. They encouraged me and through their discussions and suggestions they helped me to improve this work.

Many thanks also to Clare Davis who accepted to translate this difficult book published in French by "les Editions du Net", under the original title "Comment dessiner un mouton? Essai sur la relation genotype/phenotype" in 2015.

INDEX